Machine Learning for Operational Decisionmaking in Competition and Conflict

A Demonstration Using the Conflict in Eastern Ukraine

ERIC ROBINSON, DANIEL EGEL, GEORGE BAILEY

Prepared for the United States Army
Approved for public release; distribution is unlimited.

 ARROYO CENTER

For more information on this publication, visit **www.rand.org/t/RRA815-1**.

About RAND

The RAND Corporation is a research organization that develops solutions to public policy challenges to help make communities throughout the world safer and more secure, healthier and more prosperous. RAND is nonprofit, nonpartisan, and committed to the public interest. To learn more about RAND, visit www.rand.org.

Research Integrity

Our mission to help improve policy and decisionmaking through research and analysis is enabled through our core values of quality and objectivity and our unwavering commitment to the highest level of integrity and ethical behavior. To help ensure our research and analysis are rigorous, objective, and nonpartisan, we subject our research publications to a robust and exacting quality-assurance process; avoid both the appearance and reality of financial and other conflicts of interest through staff training, project screening, and a policy of mandatory disclosure; and pursue transparency in our research engagements through our commitment to the open publication of our research findings and recommendations, disclosure of the source of funding of published research, and policies to ensure intellectual independence. For more information, visit www.rand.org/about/research-integrity.

RAND's publications do not necessarily reflect the opinions of its research clients and sponsors.

Published by the RAND Corporation, Santa Monica, Calif.
© 2023 RAND Corporation
RAND® is a registered trademark.

Library of Congress Cataloging-in-Publication Data is available for this publication.

ISBN: 978-1-9774-1210-2

Cover: Army photo by Staff Sgt. Adriana M. Diaz-Brown.

About This Report

This report documents research and analysis conducted as part of a project entitled *Implementing Machine Learning for Assessment*, sponsored by U.S. Army Special Operations Command. The purpose of the project was to improve Army Special Operations Forces' capabilities to assess the impact and effectiveness of operations by supporting implementation of a machine learning–based assessment framework during a USASOC-identified exercise.

This research was conducted within RAND Arroyo Center's Strategy, Doctrine, and Resources Program. RAND Arroyo Center, part of the RAND Corporation, is a federally funded research and development center (FFRDC) sponsored by the United States Army.

RAND operates under a "Federal-Wide Assurance" (FWA00003425) and complies with the *Code of Federal Regulations for the Protection of Human Subjects Under United States Law* (45 CFR 46), also known as "the Common Rule," as well as with the implementation guidance set forth in DoD Instruction 3216.02. As applicable, this compliance includes reviews and approvals by RAND's Institutional Review Board (the Human Subjects Protection Committee) and by the U.S. Army. The views of sources utilized in this study are solely their own and do not represent the official policy or position of DoD or the U.S. Government.

Acknowledgments

We are indebted to the many representatives of the U.S. Army special operations community who supported this effort, although we are particularly thankful for insights from the soldiers and civilians of the U.S. Army Special Operations Command G2 (Intelligence), G5 (Plans), and G9 (Capability Development and Integration) Directorates. Specifically, we would like to thank Brooke Tannehill, Damon Cussen, LTC Tim Moore, Doug Caylor, Andy Gordon, CPT Kurtis Gruters, and Cooper Carter for their helpful insights and collaboration throughout the development of this report.

Additionally, we thank the leadership of the Strategy, Doctrine, and Resources program in RAND Arroyo Center—including Molly Dunigan, Jennifer Kavanagh (now of the Carnegie Endowment for International Peace), and Stephen Watts—for their helpful advice and guidance throughout this project's development and execution. Finally, we would like to thank John Jackson of RAND and Paul Scharre of the Center for a New American Security for their thoughtful and thorough reviews of the manuscript, which greatly improved the quality of this research.

Summary

The research reported here was completed March 2021, followed by security review by the sponsor and the U.S. Army Office of the Chief of Public Affairs, with final sign-off in June 2023.

The integration of machine learning into military decisionmaking is widely seen as critical for the United States to retain its military dominance in the 21st century. Advances in machine learning have the potential to dramatically change the character of warfare by enhancing the speed, precision, and efficacy of decisionmaking across the national security enterprise. Leaders across the U.S. Department of Defense recognize this, and a multitude of efforts are underway to effectively integrate machine learning tools across the tactical, operational, strategic, and institutional levels of war.

In this report, we explore one application of machine learning, one focused on enabling military decisionmaking at the operational level of competition and conflict. We demonstrate how machine learning, in collaboration with humans as part of a system for decisionmaking, can be used to enhance the effectiveness of military operations and activities. We show how this approach can provide commanders new insights about their operational environment through analysis of data sources that would otherwise prove inaccessible. We focus on insights that can be gained from the multitude of text-based data, such as newspaper reporting and situation reports, which are ubiquitous but are rarely integrated into decisionmaking in any systematic fashion.

The approach we describe in this report is based on the concept of a human-machine collaborative system, and we demonstrate that existing machine learning capabilities require human involvement at all stages to prove useful for operational-level decisionmaking. In this way, machine learning's development as a capability closely parallels the evolution of radar since World War II, one of the earliest examples of human-machine collaboration for military purposes. Today, much like the early warning systems used during the Battle of Britain that depended on radar machines and human observers in equal measure, machine learning still requires human involvement to direct this new sensor against the right data, interpret its output correctly, and assess the implications of its results for operational-level decisionmaking.

We demonstrate this systems approach in action using an illustrative case study, based on real-world data and a real-world crisis, placing the reader ("you") into the perspective of a military commander making notional decisions about U.S. support to the Ukrainian armed forces responding to the Russia-backed insurgency in eastern Ukraine prior to Russia's full-scale invasion in 2022. This case study is written as though you, the reader, are that commander because our goal is to emphasize the critical role you are likely to play in collaboration with machine learning tools in the future—whether as an analyst, a decisionmaker, or even as a military commander applying these tools in a similar context in the real world.

It is important to note that this case study was completed in December 2020 based on data from 2014–2020 and only analyzes conditions on the ground related to the Russia-backed insurgency in eastern Ukraine through that period. It has not been updated to reflect any insights gained since Russia's invasion of Ukraine in February 2022. And yet, this preinvasion perspective only helps to demonstrate the strengths and limitations of machine learning for operational decisionmaking in the context of real-world events that have since transpired.

Throughout the case study, you are presented the results from an actual machine learning-based assessment conducted for the purposes of this report, analyzing 18,000 historical news reports from Ukraine about the conflict from its origins in 2014 through late 2020. You will work with your staff to extract relevant insights from these data using machine learning tools and to interact with the results of this analysis to make notional decisions about the types of support to provide the Ukrainian armed forces and achieve U.S objectives in the region prior to Russia's invasion. Along the way, the strengths of human-machine collaboration learning will emerge, and you will see firsthand how machine learning tools can rapidly and systematically leverage previously inaccessible data to provide new insights on complex problems. But the limitations of this approach will emerge as well, and you will see firsthand that machine learning is only as good as the data available to support it and the human analysts who train the machine learning tool and interpret its results.

Our human-machine collaborative approach has applications across a variety of problem sets that military decisionmakers face at the operational and institutional levels of the Army and the broader Department of Defense. Thus, this research contributes to the broad literature on military applications of machine learning by demonstrating, in clear terms with concrete evidence, the trade-offs involved in using machine learning for military decisionmaking. Several key findings and recommendations for the U.S. Army emerged from this research.

Key Findings

First, our analysis demonstrates great potential for machine learning to enable military decisionmaking **but only when paired with human analysts who possess detailed understanding of the context behind a given problem.** The machine learning approach we propose here does not replace the human analyst. Rather, it makes them more efficient, more rigorous, and better able to extract insights from previously untapped sources of data. In our case study, most of the key insights gained through the use of machine learning required additional intervention from a human analyst. In some cases, it required selective layering of additional data sources on top of the model's results. In others, it required a human analyst to manually review underlying data that the machine learning tool identified as relevant and interesting. Thus, existing machine learning capabilities available to the U.S. Army will require human involvement at all stages to realize their full potential.

Second, our analysis demonstrates that **by enabling dramatic efficiency gains in the performance of repetitive tasks, a human-machine collaborative approach can analyze massive data sets at scale that would be impractical for human analysts alone, generating new insights about the operational environment that would have previously been unattainable.** Our case study demonstrates the dramatic efficiency gains that machine learning can produce, in terms of time spent by analysts on repetitive analytical tasks that involve processing large amounts of data, making analysts more efficient, more rigorous, and better able to extract insights from previously untapped sources of data. This suggests that Army leaders should prioritize machine learning as a solution to problems that require significant manual review of relevant data.

Finally, this research reveals that a systems approach to **machine learning enables standardized, objective, and long-term analysis of a multitude of data already available in an operational-level headquarters,** enhancing their potential to support effective decisionmaking. In many cases, these data are the best sources of information for making decisions at the operational and institutional levels of war, but without machine learning, the data are left to be analyzed in an ad hoc and subjective manner.

Recommendations for the U.S. Army

Two key recommendations emerged from these findings.

First, this research suggests that **the Army should provide personnel at all echelons of command with frequent exposure to machine learning** to build familiarity with how humans can leverage these capabilities as part of a military decisionmaking system.

Second, this research highlights that **the Army should build diverse machine learning teams to unlock the full potential of this capability.** These teams should integrate operations research systems analysts familiar with the details of machine learning tools, operators who have firsthand knowledge of a given operational environment, analysts who understand the data available to analyze a given problem, and commanders who can translate the machine's analysis into real-world implications for operational decisionmaking.

Contents

Figures and Tables

Figures

Tables

CHAPTER ONE

Introduction

The integration of machine learning into military decisionmaking is widely seen as critical for the United States to retain its military dominance in the 21st century.[1] Advances in machine learning have the potential to dramatically change the character of warfare by enhancing the speed, precision, and efficacy of decisionmaking across the national security enterprise. Leaders across the U.S. Department of Defense recognize this, and a multitude of efforts are underway to effectively integrate machine learning tools across the tactical, operational, strategic, and institutional levels of war.

In this report, we explore one application of machine learning, one focused on enabling military decisionmaking at the operational level of competition and conflict. We examine how machine learning, in collaboration with humans as part of a *system* for decisionmaking, can be used to enhance the effectiveness of military operations by providing commanders novel insights about their operational environments in a more efficient and rigorous manner.[2] We focus on insights that can be gained from the multi-

---

[1] Robert Work, "Remarks by Defense Deputy Secretary Robert Work at the CNAS Inaugural National Security Forum," Center for a New American Security, December 14, 2015.

[2] Work, 2015, emphasizes the importance of human-machine collaboration. This report builds on previous RAND research exploring how machine learning could be used to support operation assessment, a subset of analysis that focuses on determining whether military leaders are organized appropriately for the mission ("doing the right things") rather than whether existing tactical formations are "doing things right." See Linda Robinson, Daniel Egel, and Ryan Andrew Brown, *Measuring the Effectiveness of Special Operations*, RAND Corporation, RR-2504-A, 2019, and Daniel Egel, Ryan Andrew Brown, Linda Robinson, Mary Kate Adgie, Jasmin Léveillé, and Luke J. Matthews, *Leveraging Machine Learning for Operation Assessment*, RAND Corporation, RR-4196-A, 2022.

tude of text-based data, such as newspaper reporting and situation reports (SITREPs), which are ubiquitous but are rarely integrated into decisionmaking at the operational level in any systematic fashion.

In this context, machine learning provides a radar-like capability for the modern age, but one that can measure changes in perceptions, attitudes, and events rather than the movements of objects in space. However, current machine learning capabilities are limited in the types of insights they can provide by themselves. Thus, much like the early warning system used during the Battle of Britain, which depended on radar machines and human observers in roughly equal measure, human-machine collaboration is critical if the defense community hopes to effectively direct this new "sensor," interpret its output, and use these insights to make decisions.

Machine learning is already being used widely across the defense community. Our approach is designed to add to this existing thinking and analysis by demonstrating how integrating commanders, operators, and analysts as part of a systems approach to machine learning is critical to leveraging it for decisionmaking at the operational level of competition and conflict. We illustrate the benefits of human-machine collaboration using data from a real-world conflict, the Russia-backed insurgency in eastern Ukraine, but our approach could be adapted to a wide range of U.S. Army challenges at the operational and institutional levels of military decisionmaking.

In Chapter Two, we introduce the concept of human-machine collaboration and review recent advances in the military's use of machine learning as a system, rather than just a tool. Then, in Chapter Three, we use a detailed case study of the conflict in eastern Ukraine (prior to Russia's ensuing invasion in 2022) to demonstrate how machine learning can improve an operational-level commander's decisionmaking. This case study places you, the reader, into the perspective of a military commander, making notional decisions about U.S. support to the Ukrainian armed forces but enabled by actual results from our machine learning analysis of real-world data about the conflict prior to Russia's full-scale invasion.

In Chapter Four, we break out of this perspective to describe the strengths and limitations of this human-machine collaborative approach more generally and then, in Chapter Five, conclude with a discussion of how real-world Army users can embrace machine learning for a wide range of military applications. For those interested in our exact methodology, the technical appendix provides more detail on the mechanics of the machine learning analysis presented in our Ukraine case study.

CHAPTER TWO

Machine Learning as a System

Advances in machine learning are increasingly allowing the automation of tasks that would have previously required a human analyst. Already, these tools are providing new insights at unprecedented speeds by automating the analysis of audio, imagery, video, and other inputs. However, true artificial intelligence—the type that would allow machines to make human-like decisions and act on them autonomously—will likely remain the stuff of science fiction for the foreseeable future.[1] Instead, the real near-term innovation in machine learning is likely to come from improving the ability for humans to collaborate with machines as a team.

In this chapter, we explore the concept of human-machine collaboration and describe the idea of machine learning as a *system* to support decision-making rather than simply as a *tool* to make decisions. We begin with a brief review of recent developments in human-machine collaboration for machine learning applications across the U.S. military. We then motivate our idea of machine learning as a system by describing the development of radar during World War II, one of the earliest military applications of human-machine collaboration. We conclude with a brief discussion of machine learning applied to unstructured text data, the focus of our analysis of the conflict in Ukraine in the next chapter.

---

[1] Analysts disagree substantially on when this artificial general intelligence will first emerge, although many assess that it is unlikely within this century (e.g., Federico Berruti, Pieter Nel, and Rob Whiteman, "An Executive Primer on Artificial General Intelligence," McKinsey & Company, April 29, 2020).

Military Applications of Machine Learning and Human-Machine Collaboration

In late 2015, then-Deputy Secretary of Defense Robert Work described how technological advances in machine learning were setting the conditions for a transformation in how the United States and its adversaries competed.[2] New opportunities for "human-machine collaboration and combat teaming" allowed by machine learning stood to revolutionize intelligence analysis, providing analysts new tools to make sense from the vast volumes of data being absorbed and dramatically increasing the speed and precision of U.S. tactical weapon systems.[3]

In the years since, much of the discussion has focused on how advances in human-machine collaboration will shape capabilities at the tactical level of war and how such new capabilities will shape military strategy at the operational and strategic level.[4] Analysts have explored a multitude of ways that machine learning could enhance the lethality of U.S. forces, including using human-machine systems to defend against cyberattack,[5] deploying human-machine combat teaming (e.g., drone swarms in support of U.S. aircraft),[6]

---

[2] Work, 2015.

[3] Work described five ways in which advances in machine learning would contribute to a "third offset" strategy: (1) helping "queue intelligence systems"; (2) "human-machine collaboration, decision making" in tactical settings, such as via the F-35's innovative helmet system; (3) "assisted human operations," such as via an exoskeleton for combat troops; (4) "advanced human-machine combat teaming," which would allow interoperability of manned and unmanned systems; and (5) "network-enabled semi-autonomous weapons that are hardened to operate in an EW [electronic warfare] and cyber environment" (Work, 2015).

[4] Mara Karlin, *The Implications of Artificial Intelligence for National Security Strategy*, Brookings Institution, November 1, 2018; Frank Hoffman, "The Hypocrisy of the Techno-Moralists in the Coming Age of Autonomy," War on the Rocks, March 6, 2019.

[5] Zach Winn, "A Human-Machine Collaboration to Defend Against Cyberattacks," MIT News, February 21, 2020.

[6] Clayton Schuety and Lucas Will, "An Air Force 'Way of Swarm': Using Wargaming and Artificial Intelligence to Train Drones," War on the Rocks, September 21, 2019.

and enabling rapid analysis of the deluge of data that sensors produce.[7] Others have examined how machine learning can enable faster, more-objective, and more-precise decisionmaking in the use of lethal force,[8] improve command and control at the operational level of war,[9] enhance analysis of the information environment,[10] and facilitate planning.[11]

Many efforts are underway to effectively integrate machine learning tools across the U.S. Army and the U.S. Department of Defense in applications as diverse as vehicle maintenance (e.g., Army Stryker vehicles),[12] military recruiting,[13] and targeting.[14] Emphasis has also increased on the potential value of machine learning for the so-called institutional level of war, specifically improving the effectiveness of institutions and processes across the military that develop and deploy forces *to* the battlefield, rather than just *on* it.[15]

---

[7] Keith Dear, "A Very British AI Revolution in Intelligence Is Needed," War on the Rocks, October 19, 2018; Cheryl Pellerin, "Project Maven to Deploy Computer Algorithms to War Zone by Year's End," press release, U.S. Department of Defense, July 21, 2017.

[8] Hans Vreeland, "Targeting the Islamic State, or Why the Military Should Invest in Artificial Intelligence," War on the Rocks, May 16, 2019.

[9] Kelley M. Sayler, *Artificial Intelligence and National Security*, Congressional Research Service, R45178, August 26, 2020.

[10] Christopher Paul, Colin P. Clarke, Bonnie L. Triezenberg, David Manheim, and Bradley Wilson, *Improving C2 and Situational Awareness for Operations in and Through the Information Environment*, RAND Corporation, RR-2489-OSD, 2018.

[11] Kathleen McKendrick, "The Application of Artificial Intelligence in Operations Planning," paper presented at the 11th NATO Operations Research and Analysis (OR&A) Conference, October 9, 2017; Karel van den Bosch and Adelbert Bronkhorst, "Human-AI Cooperation to Benefit Military Decision Making," paper presented at the Big Data and Artificial Intelligence for Military Decision Making conference, Bordeaux, France, May 30–June 1, 2018.

[12] Adam Stone, "Army Logistics Integrating New AI, Cloud Capabilities," C4ISRNet, September 7, 2017.

[13] Nelson Lim, Bruce R. Orvis, and Kimberly Curry Hall, *Leveraging Big Data Analytics to Improve Military Recruiting*, RAND Corporation, RR-2621-OSD, 2019.

[14] Kelsey Atherton, "Targeting the Future of the DoD's Controversial Project Maven Initiative," C4ISRNET, July 27, 2018.

[15] For a more discussion of the institutional level of war, see Daniel Sukman, "The Institutional Level of War," Strategy Bridge, May 5, 2016.

Although often not described as such, each of these applications is similarly dependent on human-machine collaboration to be successful. This is also the case for similarly complex private-sector applications of machine learning, in which human insights derived from machine learning–curated data have frequently proven the only reliable and useful insights.[16]

Radar and the Systems Approach to Human-Machine Collaboration

Radar, which revolutionized the ability of military forces to detect enemy aircraft, was among the very first military applications of human-machine collaboration. For the British, the potency of this new technology was first demonstrated in 1940, when a network of radar stations along Britain's southern and eastern coasts gave the British a critical early warning capability during the Battle of Britain. But despite the celebrated role of this new technology in Britain's victory, early radar machines were only able to warn of aircraft flying over the English Channel. This warning would cue a group of human observers, the Observer Corps (later the Royal Observer Corps), who then provided the critical means of identifying and tracking enemy aircraft once they were inland and approaching major British cities.[17]

The development of these radar systems provides a useful framework to understand how humans learn to collaborate with machines. Early warning had evolved from individual humans making singular observations in the sky into a network of human observers connected via telephone during and

---

[16] One example of this is Watson for Oncology, a multi–billion dollar effort that was intended to automate the development of tailored recommendations for cancer treatments. In the case of Watson,

> treatment recommendations are not based on its own insights from these data. Instead, they are based exclusively on training by human overseers, who laboriously feed Watson information about how patients with specific characteristics should be treated. (Casey Ross and Ike Swetlitz, "IBM Pitched Its Watson Supercomputer as a Revolution in Cancer Care. It's Nowhere Close," STAT, September 5, 2017)

[17] Imperial War Museums, "Support from the Ground in the Battle of Britain," webpage, undated

after World War I.[18] By the beginning of World War II, these early warning systems integrated detection by radar machines into existing networks of human observers. Subsequent development of radar systems would eliminate the need for ground observers entirely, providing a comprehensive picture of the location, size, and trajectory of objects in the sky.[19] Ultimately, these systems would still require a single human operator on the receiving end of the machine, responsible for interpreting the radar screen and making rapid decisions on the fly.

Machine learning is evolving in much the same way. Until very recently, military leaders relied almost entirely on individual pieces of data, in the form of intelligence reports, battlefield outcomes, or SITREPs from subordinates, to make decisions. The field is now approaching a midpoint, not unlike radar during the Battle of Britain, in which early generation machine learning tools are now capable of rapidly analyzing large volumes of data to identify objects of interest better than humans could do on their own. However, much as during the Battle of Britain, a human analyst is still necessary to make sense of what these objects are and what they mean in all but the simplest of settings.

Machine Learning and Unstructured Text Data

This report focuses on a specific application of machine learning for analysis at the operational level of military decisionmaking: the use of machine learning to systematically and objectively extract operationally relevant insights from the large volume of unstructured text data that operational-level headquarters collect on the battlefield. The detailed textual descriptions of unfolding events captured in intelligence data, operational reporting (such as SITREPs), and traditional and social media are often the best

---

[18] Bruce D. Callandar, "The Ground Observer Corps," *Air Force Magazine,* February 1, 2006.

[19] Even as these technologies improved throughout the 1950s, human observers were still often used as a backup in the event of an enemy jamming the radar signals (Kenton Clymer, "The Ground Observer Corps: Public Relations and the Cold War in the 1950s," *Journal of Cold War Studies,* Vol. 15, No. 1, Winter 2013, p. 38).

available sources of information for many campaigns. Machine learning provides a tool for systematically incorporating these data into the commander's decisionmaking process.

Our specific application is the use of supervised machine learning to extract relevant insights from these unstructured text data. In our context, *supervised machine learning* can be described as the process of training a computer to identify when information contained in available data (e.g., a SITREP) is indicative of a particular output of interest for the assessment team (e.g., location of an enemy unit). This process relies on human analysts to manually review a subset of the available data and identify which entries are evidence of some indicator—each entry is scored with either a "1" or a "0" to indicate that it is relevant or not. The machine learning algorithm then examines the "training data" built by the human analyst, exploring the specific words used in the entries coded as relevant or not. Ultimately, the algorithm applies its own version of the coding scheme to the rest of the data in an automated fashion.

This application has clear strengths and limitations. The strength is that it allows analysis of data that would otherwise be too voluminous to include in any type of military decisionmaking process. Thus, rather than making decisions based on dozens or even hundreds of intelligence reports, this approach provides the capability to analyze thousands or potentially even millions of separate pieces of reporting in building the same decisions. Given that much of this large volume of data is likely to be irrelevant, the machine learning algorithm provides a way to rapidly sort through these data and focus the analysis on the reports that are most likely to be relevant.

The major limitation of this approach is that it requires human analysts at every step of the way. It requires human analysts to manually code the initial training data, a human to verify that the computer algorithm is performing well, and a human to make sense of these reports. The time savings from machine learning can still be dramatic, but it does not eliminate the need for a human being. Thus, machine learning is not used to magically make decisions on behalf of an analyst in a black box. Instead, it is used to help solve an existing problem more efficiently. Machine learning allows an analyst to focus on finding the right data, measuring the right things, and drawing the right conclusions rather than on crunching numbers or making educated guesses. This is the real innovation behind machine learning: less the algorithms themselves and more the new insights that humans can unleash against hard problems using previously inaccessible data.

In the next chapter, we demonstrate this concept at work through an illustrative case study that places you, the reader, into the context of a military commander tasked with making operational-level decisions informed by machine learning.

CHAPTER THREE

Demonstrating the System at Work: Machine Learning and the Conflict in Ukraine

To demonstrate this systems approach to machine learning in action, this chapter provides an illustrative case study that places you, the reader, into the perspective of a military commander making notional decisions about U.S. support to Ukrainian armed forces in their long-running fight against the Russia-backed insurgency in eastern Ukraine prior to Russia's invasion in 2022. This case study is written as if you are that commander because our goal is to emphasize the critical role you are likely to play in collaboration with machine learning tools in the future, whether as an analyst, a decision-maker, or even as a military commander applying these tools in a similar context in the real world.

In this case study, you will assume the role of a senior leader in U.S. European Command (USEUCOM), placed in charge of an operational-level headquarters in late 2020 responsible for training, advising, and equipping Ukrainian armed forces fighting against separatist and Russian aggression in eastern Ukraine. Tasked with determining the right type of support to provide to your Ukrainian partners, you will work with your staff to implement a machine learning approach that analyzes conditions on the ground in Ukraine's eastern Donbas region. Although your decisionmaking will be notional, the analysis presented in this case study is far from it. Rather, it is based on our actual application of machine learning tools to analyze a database of 18,000 Ukrainian news reports about the conflict. For those interested in the technical details of this approach, the appendix provides additional details. It is important to note that this case study was written prior to

Russia's invasion of Ukraine in February 2022 and has not been updated to reflect any insights gained since then.

We begin this chapter as you would begin your deployment as a commander, with an introduction to the operational environment in the eastern Ukraine (circa 2020) and an overview of your core objectives. Then, working primarily with the head of your intelligence directorate (G2), you will task your team to provide a systematic way of tracking several key indicators of progress toward U.S. objectives in the conflict, including levels of violence, ceasefire effectiveness, and the quality of local governance.

Armed with our database of 18,000 news reports about the conflict, your team will pilot a machine learning approach that analyzes the detailed text data available in these reports to isolate broader trends in violence, ceasefires, and governance over time. Then, your G2 will brief you with an *initial assessment* that summarizes the results from a naïve approach to machine learning, where recommendations are drawn solely from the model's results without added human context. Once you have had a chance to review these initial results, your G2 will brief a more sophisticated *systems-based assessment*, with recommendations drawn from a closer collaboration between human and machine, layering in additional data and added context to provide a more complete assessment of the conflict.

In comparing the types of insights that emerge from the initial and systems-based assessments, you will begin to understand the strengths and limitations of machine learning to enable operational decisionmaking. Foremost, you will see the critical role that human collaboration with the machine plays in producing results that are of maximum use to guide your decisionmaking. Similarly, you will glean insights into the steps required to implement a machine learning approach at scale.

We begin the case study as you take command in Ukraine.

Taking Command

It is September 2020, and you have just taken command of a newly formed operational-level headquarters in the USEUCOM area of operations that is focused on Ukraine. From this small headquarters, you report to USEUCOM leadership in Germany but now command all U.S. support to your Ukrainian counterparts in their continued fight against Russia-backed separatist forces in the country's restive Donbas region.

The conflict is now entering its seventh year. Following the country's democratic revolution in February 2014, Russian "little green men" invaded the Crimean peninsula, and pro-Russian separatists declared two independent so-called people's republics in the Donbas region along Ukraine's eastern border with Russia.[1] With overt Russian military support, the separatists have since waged a violent and sustained insurgency against the Ukrainian government.[2] Multiple international negotiations and dozens of ceasefires have failed to bring the fighting to an end after seven years of conflict.

At this stage of the conflict, your objectives in Ukraine remain focused on two broad but complementary goals: building the capacity of Ukrainian armed forces and advancing the legitimacy of the Ukrainian government's authority over the separatist regions.[3] This irregular warfare approach aims to secure Ukrainian military advantage in the conflict while buying time and space for diplomatic efforts to negotiate a successful restoration of pre-conflict Ukrainian borders.

To achieve these objectives, you have a number of tools at your disposal. Since 2016, your forces have led the Joint Multinational Training Group–Ukraine, providing training, advice, and equipment to the Ukrainian armed forces to improve their warfighting abilities. Beyond this training element, you regularly engage with senior Ukrainian counterparts on operational-level issues related to the conflict. Advisory teams from your headquarters also work with the Ukrainian Ministry of Defense to build their institutional capacity to sustain the conflict. Combined, these activities and others

---

[1] Assessing Revolutionary and Insurgent Strategies Project, *"Little Green Men": A Primer on Modern Russian Unconventional Warfare, Ukraine 2013–2014*, U.S. Special Operations Command, 2015.

[2] For a discussion of the origins of the conflict in the Donbas region, see Michael Kofman, Katya Migacheva, Brian Nichiporuk, Andrew Radin, Olesya Tkacheva, and Jenny Oberholtzer, *Lessons from Russia's Operations in Crimea and Eastern Ukraine*, RAND Corporation, RR-1498-A, 2017, pp. 43–44.

[3] For further discussion of U.S. military objectives in Ukraine, see Tod D. Wolters, "USEUCOM 2020 Posture Statement," testimony before the Senate Armed Services Committee, February 25, 2020.

seek to demonstrate the U.S. commitment to Ukraine's security and territorial integrity.[4]

Your headquarters funds these activities through the Ukraine Security Assistance Initiative, a $250 million authority to train, equip, and advise the Ukrainians in such areas as maritime situational awareness, defensive lethal capabilities (such as Javelin missiles), command and control, special operations forces, cyber defense, and even strategic communications to counter Russian misinformation.[5]

Key Conflict Indicators for Assessment

You are well aware that a successful resolution to this conflict is a difficult challenge. Given stated U.S. commitment to support the sovereignty of Ukraine, it is your job to wield the resources at your disposal to best achieve these shared objectives and make progress toward a peaceful resolution. Given your experience in the region, you decide to focus your attention on three separate but related indicators of progress on the ground in the Donbas region.

First, you know that violence continues along the line of control between Ukrainian armed forces and separatist forces in the Donbas region. Therefore, tracking larger trends in levels of violence over time could help inform the types of equipment you provide the Ukrainian armed forces, particularly if Russia is seen providing similar support to the separatists. Second, given that ceasefires have become the primary tool both sides use to deescalate tensions, you look to understand whether these ceasefires have become more or less effective over time. This could help you determine whether to build Ukrainian military capabilities more useful during a prolonged ceasefire, such as civil affairs, engineering, or intelligence. Finally, you assess that you need to know more about the quality of local governance

---

[4] For further discussion of U.S. military activities in Ukraine, see Curtis Scaparrotti, "USEUCOM 2019 Posture Statement," testimony before the Senate Armed Services Committee, March 5, 2019.

[5] U.S. Department of Defense, "DOD Announces $250M to Ukraine," press release, June 11, 2020.

inside separatist-held territory in Luhansk and Donetsk, both to gauge the popular legitimacy of the separatist governments relative to the Ukrainian state and to advise Ukrainian senior leaders on postconflict stabilization requirements.

You believe that these indicators, taken together, will help you determine the right types of equipment to provide to the Ukrainians via the Ukraine Security Assistance Initiative; the focus areas for the training and capacity-building efforts of the Joint Multinational Training Group–Ukraine and your institutional-level advisors; and the themes and messages your command should highlight in public statements, senior leader engagements with Ukrainian military officials, and interagency discussions with your U.S. counterparts from the Department of State, intelligence community, and others.

Implementing a Machine Learning Approach

You call a meeting with your G2 intelligence chief, planning to ask her to prepare a monthly report summarizing existing intelligence related to these key trends. You begin to explain your thought process, highlighting that a regular assessment of progress against these key trends will prove essential to determining how best to train, advise, and equip the Ukrainians. Once you finish giving these orders, the G2 surprises you with a counterproposal.

Rather than produce a standard intelligence assessment based on spot reports about trends on the ground, the G2 proposes to stand up a new cross-functional team to implement a machine learning–based assessment of these conflict trends. Specifically, she proposes a pilot project to demonstrate the efficacy of this approach, using publicly available historical data on the conflict to help glean lessons learned that could inform future trends in the conflict. If the analysis proves useful, she explains, the real benefit of this approach is that it could be easily updated on a monthly basis with significantly less effort from the command and with more rigor in the assessment of these indicators over time.

You remain skeptical. But, having just taken command and trying to influence a conflict now in its seventh year, you agree that it cannot hurt to try a fresh approach. As a result, later that day, you direct the command to stand up a cross-functional team to support this effort, with intelligence analysts, operators, planners, and strategists, as well as an operations research systems analyst (ORSA) drawn from the command's staff sections to round out the team's expertise. You task the G2 to report back within a week.

Gathering the Right Data

The next day, the new machine learning cross-functional team meets for the first time. The various representatives from each of the command's staff sections discuss the three indicators you would like to assess over time— levels of violence, ceasefire effectiveness, and quality of local governance— and which data sources could be used to provide a historical look at these trends suitable for machine learning analysis. One analyst notes that, if this pilot proves successful, whatever data source the team chooses should, ideally, be updated regularly going forward. Ultimately, the team decides that the Ukrainian National News Agency (UKRINFORM), a reputable news source, is likely to provide a comprehensive source of consistent reporting about the conflict dating back to 2014.[6]

Using a commercial database, the team's ORSA then downloads every news report from UKRINFORM that mentions Donetsk, Luhansk, and eastern Ukraine, hoping to isolate just the news reports most likely to contain relevant information about the conflict. The result is a rich library of nearly 18,000 separate news articles describing the conflict since 2014, summarized in Figure 3.1.

Training the Machine Learning Model to Read the News

The machine learning team contemplates reading every one of these 18,000 news reports, manually looking for evidence of violence, ceasefires, and

---

[6] UKRINFORM is the national news service of Ukraine. Founded in 1918, UKRINFORM has bureaus throughout Ukraine and in ten other countries. It publishes news articles in Ukrainian, Russian, English, and six other languages. The team restricted their analysis to English language news only, which could induce some bias if UKRINFORM were to publish certain aspects of the conflict in Ukrainian but not in English. Further analysis could use machine learning to either incorporate machine-translated Ukrainian language news into this analysis or focus solely on Ukrainian language news.

FIGURE 3.1

Ukrainian Media Reports Mentioning Donetsk, Luhansk, or Eastern Ukraine

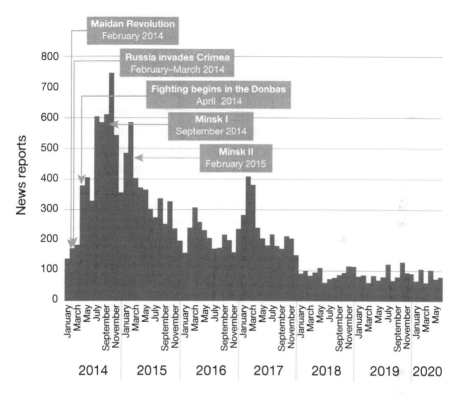

SOURCE: UKRINFORM news reports accessed through the Nexis-Uni academic research tool.

governance over time. But they quickly decide that this approach would take far too long and that the team would almost certainly struggle to code these data in an objective and reliable manner across teammates and over time. Per your initial directive, they begin to implement a machine learning approach to read through these news reports more efficiently.

First, the team reads through a small, random sample of 300 UKRIN-FORM news articles drawn from the nearly 18,000 total reports in the data set. Based on guidance from the team's intelligence analyst, team members look for evidence of specific events related to your three conflict indicators of interest, including attacks by separatist forces, shelling of civilians, Rus-

The G2 starts the briefing with a discussion of her team's machine learning analysis of insurgent violence in the Donbas region.

Levels of Violence and Russian Involvement

To understand how levels of violence have changed over time since the start of the conflict, the G2 explains that the team read an initial batch of UKRINFORM news reports (the training data) to code those mentioning specific violent incidents in in the Donbas region. Similarly, team members flagged the news reports that mentioned direct Russian involvement in that violence or general Russian military support to separatist forces. Using these training data, they then trained a machine learning tool to look for similar violent incidents in the rest of the 18,000 UKRINFORM news reports.

On a slide, they show you the results of this analysis, plotting trends in violence and Russian support to violence over time (Figure 3.2).[11] This chart shows the number of news reports over time for which the machine learning model was reasonably confident that the report referenced a specific violent incident in the Donbas region, or Russian involvement in such violence.

Initial Assessment

Based solely on these data, the G2 suggests that the levels of violence in Donetsk and Luhansk have been remarkably low since 2018, particularly in comparison with the initial stages of the conflict. Although periodic flare-ups have occurred in recent years (such as in summer 2019 and early 2020), these spikes in violence appear incredibly small in comparison to the earlier periods of more-active fighting. Similarly, this analysis suggests that Russian support to the separatists, or direct involvement of Russian forces in the conflict, appears to have been most intense during the war's most violent periods, in late 2014 and 2015, as well as early 2017. However, the G2 explains that its machine learning analysis of news reports finds relatively minimal Russian support to the separatists in recent years.

To offer some added perspective, the head of your operations directorate (G3) jumps in to discuss the implications of these findings. At face value, he

---

[11] For a full discussion of coding criteria, training data, model calibration, and visualization of results, see the technical appendix to this report.

FIGURE 3.2

Machine Learning Results for Levels of Violence and Russian Involvement

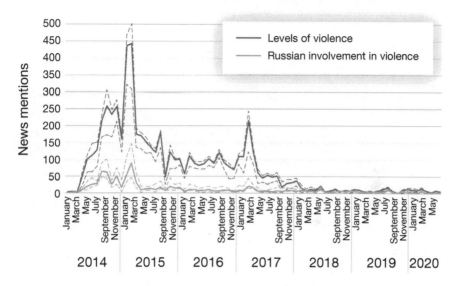

SOURCE: Authors' calculations based on news reporting from UKRINFORM news reports accessed through the Nexis-Uni academic research tool.

NOTE: Dashed error bands are shown around each line, demonstrating alternative thresholds above and below the chosen likelihood thresholds that the machine learning algorithm produced. The technical appendix provides details on the exact thresholds chosen.

explains, these results suggest that you should direct the Joint Multinational Training Group–Ukraine to focus less on preparing Ukrainian troops for immediate combat and more on equipping the force for long-term growth and professionalization. Specifically, these results suggest that you could consider lengthening training timelines because Ukrainian forces are not likely to be needed on the front lines of the conflict on the conclusion of their training. This training should therefore focus less on small-unit tactics useful for skirmishes with separatists and more on advanced warfighting capabilities that are likely to deter Russian escalation of the conflict as seen in its early years.

Systems-Based Assessment

But you challenge the machine learning team on these findings. Through other sources of data, such as the SITREPs that you receive from your subordinate commanders in the field, you know that fighting along the line of control in eastern Ukraine has remained relatively prevalent in recent months. You look back to your G2, who actually appears to share your skepticism.

When pressed, she explains that the apparent decline in violence in recent years shown in this initial machine learning assessment may be, to a great extent, the result of a major decline in the overall volume of news reporting about the conflict from UKRINFORM over the same time period. Thus, it is possible in this instance that the result of the initial analysis is being skewed by reporting biases in the data rather than true changes in degree of violence. To give you a better sense of actual levels of violence in the conflict, the G2 suggests that there are two additional courses of action to layer added context on top of the machine learning model's analysis of trends in these news reports.

First, per your suggestion up front, the G2 suggests using the same machine learning approach to analyze the SITREPs you receive every day from your subordinate commanders. These data are likely to contain more granular information on insurgent attacks than are available in news reports. At this suggestion, your command's ORSA jumps into the discussion, noting that a machine learning tool could prove particularly useful at extracting relevant trends from these data, since they are full of large amounts of unstructured text in the same way as news reports.

Additionally, the G2 proposes to validate the decline in violence seen in this analysis using two other publicly available data sources on violence in the Donbas region, Janes Information Services' Terrorism and Insurgency Database, and the Armed Conflict Location Event Dataset (ACLED).[12] She then directs you to the backup slides of your briefing, highlighting that both Janes and ACLED data demonstrate a decline in levels of violence in the

---

[12] Janes Information Services is a subscription-based service that, among other services, collects data on attacks by nonstate armed groups in a number of different conflict zones. ACLED is a scholarly database of similar attacks that is publicly available. Additional discussion of these data is available in the technical appendix to this report.

Donbas region since 2018, but not nearly as drastic a decline as the machine learning results show.[13]

Given this added context, your G3 operations chief suggests that it would be premature to reorient training away from skills most relevant to current combat skills needed along the frontlines. However, he recommends that if violence continues to decline throughout the rest of the year, perhaps indicated by a new round of machine learning analysis of your SITREPs, you should consider reevaluating your approach.

You then press the team about whether a more sophisticated analysis would reveal anything different about the model's initial characterization of Russian involvement in violence over time. Even though the volume of news reporting on the conflict has declined in recent years, the G2 highlights that a basic statistical analysis of the machine learning results reveals that levels of Russian involvement in violence are more closely correlated with levels of violence during the most active periods of fighting rather than periods of relative stability.[14] This suggests, but does not prove definitively, that escalation in Russian support to the separatists may be a driving factor behind prior overall escalations of the conflict. Thankfully, the news reports that the machine learning team used appear to provide sufficient granularity to tease out this effect. The G2 suggests that, if this effect holds true going forward, these results mean you should prioritize indicators and warning of potential movements of Russian forces near the border as a harbinger of future escalations of fighting along the line of control.

---

[13] Janes, for example, identified 1,505 separate attacks by nonstate armed groups in Ukraine in the first half of 2020. If these trends were to continue throughout the second half of the year, these estimates suggest that 2020 would be only 33 percent less violent than in 2016. Further analysis of Janes' and ACLED's data is available in the technical appendix. See Janes Information Services, "Terrorism and Insurgency Database," 2020.

[14] During years with escalating conflict (in 2014, 2015, and 2017), the correlation coefficients between levels of violence and Russian involvement month over month were 0.878, 0.91, and 0.853, respectively. During years with stable or declining levels of violence (in 2016, 2018, 2019, and the first half of 2020), the correlation coefficients between levels of violence and Russian involvement month-over-month were 0.473, 0.746, 0.657, and 0.776, respectively.

Assessing Ceasefire Effectiveness

The briefing then turns to your second priority indicator of conflict trends in eastern Ukraine—whether ceasefires have become more or less effective over time. Your command's political advisor (POLAD), a civilian Department of State employee assigned to your headquarters to advise you on Ukrainian political dynamics, explains that ceasefires have become a frequent tool that the various sides of the conflict use to de-escalate tensions, dating back to initial rounds of negotiations in Minsk in October 2014 and February 2015. Similarly, as the POLAD explains, the international community has relied on such ceasefires to buy time and build support for a diplomatic solution to the conflict. Although these ceasefires have been frequent, few overall have proven lasting. In all, the POLAD notes that nearly two dozen ceasefire attempts have been attempted, at least based on Russian estimates in late 2018.[15]

The G2 flips to the next slide of your briefing and shows a graph plotting trends in the effectiveness of ceasefires over time, as flagged by the machine learning model's analysis of UKRINFORM news reports (Figure 3.3). This analysis, she explains, flagged news reports highlighting any evidence of successful ceasefires (including the cessation of violence or withdrawal of forces) or failed ceasefires (including the violation of specific terms of any given ceasefire).[16]

Initial Assessment

The G2 highlights two noteworthy trends from this initial naïve analysis based solely on the machine learning results, focusing only on the trends that may affect the types of training and equipment you provide to your Ukrainian partners while in command. The first is that a ceasefire attempt in September 2015, although not the first such ceasefire, seems to have been the most successful; it produced a major spike in news reports discussing the

---

[15] "New Year Ceasefire Enters into Force in Donbass," TASS Russian News Agency, December 28, 2018.

[16] The technical appendix provides additional details regarding our coding criteria.

FIGURE 3.3

Machine Learning Results for Ceasefire Effectiveness

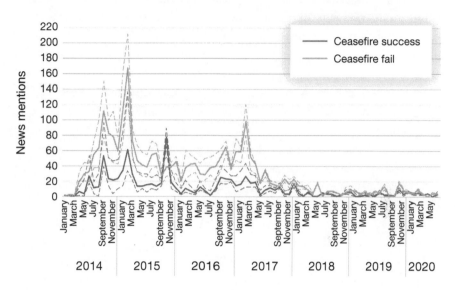

SOURCE: Authors' calculations based on news reporting from UKRINFORM news reports accessed through the Nexis-Uni academic research tool.

NOTE: Dashed error bands are shown around each line, demonstrating alternative thresholds above and below the chosen likelihood thresholds produced by the machine learning algorithm. The technical appendix provides details on the exact thresholds chosen.

withdrawal of separatist forces from the front lines.[17] This spike occurred without any comparable increase in evidence of ceasefire violations, suggesting either brief progress along the front lines or, at a minimum, a brief spike in public optimism about ceasefires as reported in the news. Perhaps, she explains, lessons can be learned from this specific ceasefire that can be replicated in the future. However, she is quick to note that this ceasefire lasted only three days, despite the high volume of news reports discussing its effectiveness.

Alternatively, the G2 notes that the machine learning tool appears to flag a more sustained uptick in ceasefire adherence for several months following

---

[17] For a discussion of this ceasefire, see Thomas Gibbons-Neff, "Three-Day-Old Ceasefire in Ukraine Broken as Fighting Resumes in Some Areas," *Washington Post*, September 3, 2015.

an October 2016 summit between Russian, Ukrainian, German, and French leaders. But, unlike the first peak in ceasefire success from September 2015, the G2 notes that this ceasefire attempt was accompanied by an even greater spike in evidence of ceasefire violations over the same period. She notes that similar trends have continued since, with upticks in evidence of successful ceasefires occurring roughly simultaneous with increased evidence of ceasefire failures through early 2020. Unfortunately, there appears to be significantly less discussion in Ukrainian news since 2018 about these ceasefires, either successful or failed, at least when compared with earlier periods of the conflict.

Broadly speaking, the G2 suggests that these results mean you should remain skeptical of the potential for major changes to come on the battlefield from any one new ceasefire or, at least, changes serious enough that you should adjust the types of assistance the command provides to the Ukrainian armed forces as a result. The G3 operations chief adds that as you engage with your counterparts from across the U.S. government on potential new ceasefires, as well as with the Ukrainian Ministry of Defence, these results suggest that you should continue to emphasize the importance of battlefield preparation and training even during the initial stages of any new ceasefire. After all, the military training that you provide to the Ukrainian armed forces is intended to buy time and space for diplomacy to succeed.

Systems-Based Assessment

Pleased with these results but skeptical that there may be more to the story, you ask the machine learning team if there are other sources of data that can validate these findings. In particular, you highlight that you regularly review daily spot reports from the Organization for Security Co-operation in Europe (OSCE) that catalogue ceasefire violations by various actors in the conflict over time. These reports, drawn from the observations of teams of OSCE personnel on the ground in the Donbas region, provide an internationally recognized accountability mechanism that could be helpful to integrate into this analysis.[18]

---

[18] For data from 2016–2019, see OSCE Special Monitoring Mission (SMM) to Ukraine, "2019 Trends and Observations," webpage, 2019. For data from the first quarter of 2020, see OSCE SMM Ukraine, "Trends and Observations: Jan–Mar 2020," webpage, 2020a.

The machine learning team's ORSA jumps in to respond, noting that the OSCE's spot reports are often already included in the underlying UKRINFORM news data the machine learning model analyzes. However, the ORSA explains, the machine learning approach used in this case is only able to identify *whether* a given OSCE report cites evidence of a ceasefire violation, not *how many* violations are mentioned in the report. To get a sense of potential differences between the machine learning approach and the OSCE data on their own, the ORSA then shows you a graph summarizing the broader trends in ceasefire violations captured in the OSCE's monitoring reports (Figure 3.4).

You see that the OSCE data clearly affirm that ceasefire violations have indeed declined over time, as reported in the machine learning analysis. However, it is clear that that, as before with the analysis of levels of violence, this decline is not nearly as substantial as in the machine learning analysis. In fact, ceasefire violations appear to remain a relatively prominent fixture of the conflict.

FIGURE 3.4

OSCE Ceasefire Violations in Ukraine, 2016–2020

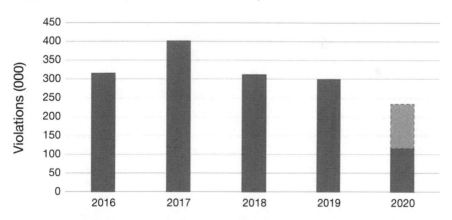

SOURCE: OSCE SMM Ukraine, 2019; OSCE SMM Ukraine, 2020a; OSCE SMM Ukraine, 2020b.
NOTE: Data for the second half of 2020 are imputed and shown in green for illustrative purposes only. This imputation assumes an identical number of ceasefire violations will occur between July and December 2020 as occurred between January and June.

For data from the second quarter of 2020, see OSCE SMM Ukraine, "Trends and Observations: Apr–Jun 2020," webpage, 2020b.

When pressed to dig into these data further, the ORSA explains that he was able to extract additional information from the OSCE's spot reports on the *types* of ceasefire violations that have occurred over time. These data, he explains, may prove particularly useful for the more-recent years, for which the machine learning analysis lacked significant volume of data in the underlying set of UKRINFORM reports. As the OSCE describes in each of its reports, ceasefire violations are aggregated across a number of different events, such as live-fire exercises, explosions from indirect fire, failure to withdraw weapons from agreed-on lines, and efforts to restrict the freedom of movement of observers. Specifically, it appears that there has been a recent increase in the volume of ceasefire violations caused by indirect fire along the front lines, particularly in the first half of 2020.[19] By comparison, the number of ceasefire violations from other factors, such as failure to withdraw banned weaponry, has fallen considerably over the same period. Similarly, separatists' efforts to restrict the freedom of movement of observers were only marginally higher in early 2020 than in prior years.

You ask your team whether the OSCE data, or other sources of information, can also speak to the frequency with which the separatist forces and Russian forces have successfully adhered to ceasefire conditions. As the initial machine learning analysis demonstrates, the absence of ceasefire violations does not necessarily equate to successful ceasefire adherence. The G2 notes that there are few, if any, publicly available alternatives to news reports in the public domain that catalogue direct evidence of Russian involvement.

You conclude this discussion by asking for the team's assessment of the implications of this analysis, particularly in light of their efforts to layer in additional data on top of the machine learning results. The G3 operations chief answers quickly, suggesting that the added results do not change the team's recommendations substantively. However, he notes, this does suggest that helping the Ukrainians defend against indirect fire may be a prudent focus area.

---

[19] In 2019, OSCE observers flagged 3,373 incidents of explosions from indirect fire (specifically those from the Multiple Launch Rocket System, as well as other artillery, mortars, and tanks). Through the first six months of 2020, OSCE observers flagged 3,010 incidents of the same type. If these trends were to carry forward through the rest of the year, this would represent a 78-percent increase year over year.

To refine these recommendations in the future, the team proposes to supplement its existing body of news reports from UKRINFORM with additional Ukrainian or Russian-language news sources that provide more discussion about conditions along the line of control. Given the usefulness of the OSCE data, the team believes added context can improve the machine learning analysis to give you better insights going forward.

Assessing the Quality of Local Governance

The G2 then transitions to the final portion of the briefing, focused on the team's machine learning–based analysis of the quality of local governance inside the separatist regions. The G2 begins with some context, highlighting that the leaders of the self-declared Donetsk People's Republic (DPR) and Luhansk People's Republic (LPR) have established semiautonomous proto-states, with constitutions, government agencies, security and police forces, schools, and public services.[20] She also highlights that Russia has consistently provided humanitarian and economic support to these regions, and advocated that Ukraine codify the de facto devolution of power to the separatists.[21]

The G2 explains that the team's machine learning analysis attempts to assess whether these Russia-backed separatist governments have effectively governed these territories. Similarly, this analysis should provide insights into the intent of Russia and the separatist governments themselves. Like many rebel groups, the DPR and LPR may provide public services to bolster the legitimacy of their claims over the Donbas region. However, these groups face a trade-off between devoting resources to effective governance

---

[20] Significant debate exists about whether these governments are legitimate representatives of local popular will within the Donbas region or whether they are simply manifestations of Moscow's direct political influence. See for example, Sabine Fischer, *The Donbas Conflict: Opposing Interests and Narratives, Difficult Peace Process*, Berlin: German Institute for International and Security Affairs, April 2019.

[21] Russia has advocated that Ukraine adopt a federalized system that would allow the pro-Russian Donbas region to effectively veto any decision by Kyiv to move closer to the European Union or NATO. For further discussion, see Paul Niland, "Making Sense of Minsk: Decentralization, Special Status, and Federalism," Atlantic Council, January 27, 2016.

and devoting them to conducting attacks against Ukrainian armed forces.[22] Moreover, increased violence on the battlefield is likely to damage infrastructure and harm separatist efforts to govern their territories effectively. Given these trade-offs, clear evidence of successful separatist governance could demonstrate that the separatists seek to transition toward peace rather than to continue the conflict through violence.

The G2 then briefs the results of the machine learning analysis of UKRINFORM news reports. She explains that the model was trained to look for evidence of effective governance (shown in blue in Figure 3.5), including the successful provision of public services by the separatist governments, along with evidence of humanitarian or economic support provided by Russia. The machine learning tool was also trained to look for evidence of ineffective governance (shown in green), including evidence of poverty or deprivation in the separatist regions, failed public services, and damage to infrastructure caused by separatist violence.

Initial Assessment

The G2 notes that, taking these results on their own, two trends emerge from this analysis that are relevant to local governance. First, these results suggest that neither the separatists nor their Russian backers have devoted significant effort to govern the Donbas region over time, even as the conflict has matured. In particular, although Russia's use of so-called humanitarian convoys to provide support to the Donbas region has become a major point of contention between the two sides, the machine identifies little evidence that this tactic has become more prevalent over time.[23] Either way, the G2 notes that there is little in this analysis to suggest that Russia and the separatists have made marked improvements in local governance to help these

---

[22] For a broader discussion of the resource trade-offs between rebel governance and violence, see Adrian Florea, "Rebel Governance in De Facto States," *European Journal of International Relations*, Vol. 26, No. 4, December 2020, p. 1007.

[23] The team is careful to explain that the exact balance between military and humanitarian aid in these shipments is not readily known, given that Russia refuses to allow inspections by Ukrainian customs agents or OSCE observers. For a recent example of a Russian humanitarian convoy into the disputed territories, see "Russia Sends 98th Humanitarian Convoy to Donbas, Ukraine's Foreign Ministry Protests," 112 Ukraine, August 28, 2020.

FIGURE 3.5

Machine Learning Results for Quality of Local Governance

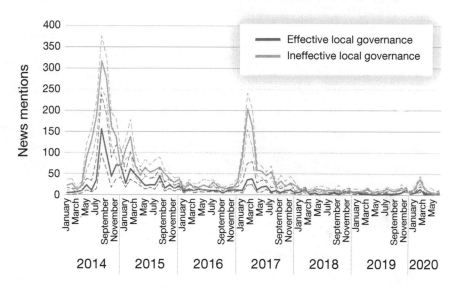

SOURCE: Authors' calculations based on news reporting from UKRINFORM news reports accessed through the Nexis-Uni academic research tool.

NOTE: Dashed error bands are shown around each line, demonstrating alternative thresholds above and below the chosen likelihood thresholds produced by the machine learning algorithm. The technical appendix provides details on the exact thresholds chosen.

regions transition from protostates into more autonomous regions capable of self-governing.

Second, more generally, the team's ORSA adds that the quality of local governance in the separatist regions actually received little attention in Ukrainian news during the first few years of the conflict through 2017, particularly when compared with the significant number of news reports discussing ongoing fighting and ceasefire effectiveness over the same period. With the exception of the few periods of most intense fighting in which significant infrastructure was destroyed, relatively few news articles mention local service provision, economic conditions, or other quality-of-life indicators in 2015 and 2016.[24] The team's ORSA flags that, in more recent months, such as March 2019 and February 2020, there have actually been modest upticks in discussion of failed local governance. But at face value, these upticks appear inconsequential.

As commander, your G2 recommends that the only actionable insights you should glean from this analysis are related to the larger intent of the Russian and separatist regions. The lack of evidence that the separatist governments are prioritizing high quality of governance over time suggests that they remain focused on the military aspects of the ongoing conflict rather than transitioning toward state-like governance independent of the Ukrainian state. If true, this would imply that your training and equipping efforts should remain focused on combat preparedness rather than on helping Ukrainian forces stabilize conflict-affected regions.

And yet, the G2 caveats, the overall lack of discussion of governance in UKRINFORM news reports makes these findings tentative at best.

Systems-Based Assessment

You ask the machine learning team if there are other sources of data that could be used to expand on these results. The G2 responds quickly, noting that, unlike the other indicators discussed so far, there are relatively few alternative sources of data about governance and service provision inside the separatist-held territory in eastern Ukraine. In fact, your machine learning team suggests that this approach provides a novel, empirical look into the *quality* of governance that would be difficult to replicate through other data sources.[25] So, rather than layer additional data on top, the G2 suggests

---

[24] This is likely a product of increased *demand* for news articles about the conflict rather than a reduced *supply* of newsworthy stories about quality of life inside the Donbas region—a limitation of using news reports to measure governance.

[25] One approach could be to conduct survey research inside the separatist-held territory; however, this is often infeasible. An alternative approach could be to use satellite imagery and remote sensing data to measure economic conditions as a proxy for quality of governance. For an example of this approach, see Eric Robinson, Daniel Egel, Patrick Johnston, Sean Mann, Alex Rothenberg, and David Stebbins, *When the Islamic State Comes to Town: The Economic Impact of Islamic State Governance in Iraq and Syria*, RAND Corporation, RR-1970-RC, 2017.

that more insight can be gleaned by digging further into the underlying news reports analyzed by the machine learning tool.

As it turns out, the team's ORSA explains, a more thorough review of news reports about local governance in the Donbas region reveals something interesting. A review of a subset of these reports indicated that initial upticks in failed governance in early years of the conflict appear primarily to have been due to news reports about damage to critical infrastructure caused by separatist violence. Conversely, more-recent upticks in news reports about failed governance primarily reference the political and logistical hurdles to reintegration of the separatist regions into a sovereign Ukraine.[26] As an example, the ORSA notes that upticks in March 2019 and February 2020 were the result of a variety of news stories focused on the decision to cut off power and water supplies to portions of Donetsk and Luhansk because of their failure to pay debts to Ukrainian public service providers from earlier in the conflict.[27]

So, what does this mean for you as the commander responsible for U.S. support to Ukraine? First, these results suggest that the level of infrastructure damage inside the separatist-held territories in Donetsk and Luhansk could be declining. Your G3 operations chief explains that, if the machine learning analysis were to flag a future spike in damage to local infrastructure in news reports, you could direct the Joint Multinational Training

---

[26] Although the volume of news mentions to this effect in these months may appear small, they actually represented a significant volume relative to the total number of news articles published by UKRINFORM about the Donbas region in each of these months. Our machine learning analysis suggests that, across the 61 articles written about the Donbas region in March 2019, there were 15 unique mentions of some aspect of failed local governance (one mention per every 4.1 articles). By comparison, there were only four unique mentions of violent incidents in that month (one mention per every 15 articles). In February 2020, there were 38 unique mentions of failed local governance out of 105 separate news reports (one mention per every 2.8 articles). By comparison, there were only 13 unique mentions of violent incidents that same month (one mention per every 8.1 articles).

[27] See, for example, "Luhansk Energy Association to De-Energize Popasnianskyi District Water Supply Channel Supplying with Water 75,000 Residents of Luhansk Region on February 21 Due to UAH 24.3 Million Debts," Ukrainian News Agency, February 17, 2020.

Group–Ukraine to train and equip Ukrainian military engineers to assist with the restoration of critical infrastructure.

The G2 notes that, if future machine learning analysis of news reports were to flag a change toward greater Russian support to separatist governance, this could signal a broader change in Russian strategic intent toward building an independent Donetsk and Luhansk. Similarly, your G3 operations chief suggests that you could then advocate for Ukraine Security Assistance Initiative funding to train and equip a Ukrainian military civil affairs capability in response. Your POLAD also suggests that you could work with your interagency counterparts, such as the U.S. Agency for International Development, to increase targeted economic assistance to Ukrainian-controlled regions in Donetsk and Luhansk just across the line of contact.

For now, however, your G2 explains that that both a face-value reading of the machine learning results and her team's more thorough review of the underlying data suggest neither of these courses of action is urgently required.[28]

Way Ahead

The G2 concludes the briefing, highlighting that this machine learning pilot should be seen as an opening salvo, intended to demonstrate how machine learning could be used to support your operational-level decisionmaking while in command. The team, she explains, focused on a single source of news reporting at first and, where gaps in the analysis remained, demonstrated the potential to layer in additional data sets and human context that transformed machine learning from a decisionmaking *tool* to a decision-making *system*.

---

[28] Of note, academic research into DPR and LPR provision of governance affirms these findings, suggesting that the separatists and their Russian external sponsors have made little effort to promote effective local governance. For example, one study notes that the "LPR has its own government, parliament, and courts, but these institutions are hardly functional: the government is riven by rivalry among former rebel commanders, the parliament is little more than a rubber-stamp assembly that convenes irregularly, and the newly established local courts have yet to adopt a civil code" (Florea, 2020, p. 6).

Advocating to continue this approach at scale, the G2 explains that, while you may have been briefed similar conclusions about Ukraine in the past, her machine learning cross-functional team came to many of the same conclusions using an objective analysis of trends that took a relatively modest amount of effort to produce. Ultimately, the model's analysis proved useful in its own right in some cases but also helped your command prioritize the right sources of amplifying data to gain a more complete understanding of the conflict. Overall, as the G2 explains, the team spent less time crunching data and more time focusing on their implications for your command and on alternative explanations of the trends the machine learning model revealed. Given scarce resources, these efficiency gains could prove useful to you as a commander.

However, as you pointed out periodically throughout the team's presentation of its analysis, there are clear limitations to relying solely on machine learning to make decisions at face value. After all, this approach is not meant to provide a standalone decision tool to replace the combined insight of you as the commander along with operators, intelligence analysts, planners, and other advisors within your command. Rather, this case study demonstrates that a deliberate systems-based approach to machine learning can produce more-objective results and can foster sustained collaboration between you, your staff, the machine learning tool, and a variety of sources of other data.

With that, the G2 concludes the briefing, pending your questions.

CHAPTER FOUR

Strengths and Limitations of Machine Learning

Chapter Three presented a case study of machine learning in action, in the context of notional decisionmaking by an operational commander in Ukraine. This case study revealed that, as with early radar systems at the beginning of World War II, machine learning can be effective at flagging relevant trends based on observable data. At the same time, this case study makes clear that machine learning is not yet prepared to replace the human analyst in this context. Rather, it makes these analysts more efficient, more productive, and more rigorous by working in tandem with machine learning's ability to analyze large volumes of data. Ultimately, machine learning frees humans to spend more time on sense-making and less time crunching numbers. But this kind of machine learning approach has some major limitations, the foremost being that the insights from this approach are only as good as the underlying data and the analysts assigned to prepare and analyze the data.

In this chapter, we describe these strengths and limitations in the context of our analysis of the conflict in eastern Ukraine in Chapter Three.

Machine Learning Produces Significant Efficiency Gains

While producing our machine learning analysis of the conflict in Ukraine, we carefully catalogued the amount of time spent during each stage of the analysis. Using these numbers, we calculated the precise time that the use of a machine learning approach to code unstructured text data saved over a manual approach. Table 4.1 presents these estimates in greater detail. We report the number of hours of labor spent on each phase of the machine learning analysis and, using the estimates, calculate the amount of time it would take to replicate this analysis entirely by hand. We estimate that a machine learning approach was nearly 40 percent more efficient overall than producing this analysis by hand, assuming that the machine learning

TABLE 4.1

Efficiency Gains from Machine Learning

		Hours of Labor	
		Machine Learning	Manual Approach
Phase One Write code	Data preparation	10.0	0.0
	Machine learning algorithm	15.0	0.0
	Visualize results	5.0	0.0
Phase Two Code historical news reports (since 2014)	Prepare the data • Download 3,600 articles per hour x 18,000 articles	5.0	5.0
	Code data • Machine learning: 30 minutes (for 400 extracts) x 3 indicators x 3 calibrations • Manual: 3.5 articles per minute	4.5	86.0
	Calibrate algorithm • 1 hour x 3 indicators x 3 calibrations	9.0	0.0
Phase Three Code monthly updates (for a year)	Prepare the data • Download 3,600 articles per hour x 100 articles x 12 months	0.5	0.5
	Code data • Machine learning: 1 minute (for 10 extracts) x 3 indicators x 12 months Manual: 3.5 articles per minute	0.5	5.5
	Calibrate algorithm • 1 hour x 3 indicators x 1 calibration x 3 times per year	9.0	0.0
Total time required		58.5	97.0

NOTE: Estimates are provided in hours, rounded to the closest half hour.

team continues to update its analysis of Ukraine monthly over the course of a year.

But this aggregate measure actually understates the real efficiency gains from machine learning, particularly in the context of the individual steps required to produce this analysis. Our machine learning approach took a total of just under 60 hours to analyze the 17,995 news articles in our sample. This included a substantial up-front cost, taking an estimated 30 hours to write the computer code required to extract the news data into a machine-readable format, code the machine learning algorithm, and write the code required to eventually visualize the results (as seen in the figures in this report).

We then spent roughly 13.5 hours to code three rounds of training data and calibrate our machine learning algorithm, reviewing 900 extracts from these news reports for each of our three indicators.[1] We estimate that it would take at least 86 hours for a human analyst to review and manually code the entirety of the 17,995 reports. This suggests that machine learning is at least six times more efficient than a manual approach in terms of the time an analyst spends on repetitive tasks analyzing data to produce this analysis.

Efficiency Gains Depend on the Volume of Data

This analysis also suggests that machine learning becomes less efficient than manual coding for smaller amounts of data. In phase three of this table, we estimated the amount of time required to update this analysis on a monthly basis going forward. Based on the volume of UKRINFORM reporting about the conflict in recent years, we assumed that only 100 additional news reports are likely to be published each month going forward, a significant reduction from the heavier volume of news published about the conflict in its early years. We estimate that, over the course of a year, it would take 10 hours of labor for a machine learning approach (including coding of train-

---

[1] Our machine learning approach relies on three-sentence extracts drawn from each news report, identified using relevant keywords for each indicator. A human analyst would instead be required to read the entirety of each news report, which is part of the added cost required to replicate this analysis manually.

ing data and recalibration of the model), but only 6 hours of labor for a human to simply read all the reports manually. Although this time savings is hardly significant in practical terms, it does make clear that, when smaller amounts of data are available, time may be better spent diving into the data manually rather than calibrating a machine.

However, the inverse is also true: Machine learning becomes much more efficient than manual coding for larger amounts of data. If we assume that 10,000 news reports had to be analyzed every additional month (rather than 100), it would take the human roughly 600 hours to read through that volume of news reports over a year. Our experience suggests that a machine learning approach would be ten times more efficient, taking only 60 hours of labor to code training data and calibrate the machine learning tool to produce a similar analysis.[2]

Machine Learning Opens New Lines of Inquiry

This raises perhaps the most compelling reason to embrace machine learning for operational decisionmaking. The real benefit of the machine learning approach we used to analyze the conflict in Ukraine is that we attempted the analysis in the first place. Without machine learning as an option, the 86 hours of time required to manually code all 18,000 news reports would almost certainly be prohibitive for the staff of a busy operational-level headquarters. The headquarters would be left without the insights generated from news data or the added context that comes from layering secondary sources of data on top of the machine learning results

Machine learning therefore creates more than just time savings—it opens up new lines of inquiry that leverage new data sources, allowing

---

[2]   These estimates are based on the assumptions in Table 4.1. We assume that it would take 33.5 hours to download 10,000 articles a month for 12 months in both cases (machine and manual). We then estimate that it would take roughly 18 hours to code training data for machine learning over the course of a year, compared to 571 hours to manually read the total 120,000 news reports. Model calibration, as before, would require an additional 9 hours for the machine learning approach. Together, these estimates suggest 60 hours of time for machine learning and 605 hours of time for manual analysis.

staff to spend less time crunching numbers and more time making better-informed decisions. Although news reports proved only somewhat useful in this case study, other sources of text data available to an operational-level commander about a conflict could prove significantly more important. These include SITREPs, intelligence reports, and such other forms of open-source reporting as social media. As it stands, these other forms of unstructured text data are episodically analyzed by decisionmakers, primarily one at a time and at varied intervals over time. Machine learning could help an operational-level command systematically, rigorously, and objectively analyze these data for new insights.

Efficiency Comes at the Expense of Accuracy

However, it is possible that the time savings and new lines of inquiry from a machine learning approach come at the expense of accuracy. After all, in the case of the Ukraine case study presented in the last chapter, our notional military headquarters relies on machine-coded analysis of news reports that could be prone to misdiagnosing certain events over others. How confident are we that the machine learning tool was accurately reading the news? The only way to know for sure is to manually code a sample of data already analyzed by the machine learning tool and compare the results.

To do this comparison, we examined the training data used for the machine learning analysis in our Ukraine case study in greater detail.

Over the course of two successive rounds of model calibration, we fed a machine learning tool roughly 700 manually coded news reports intended to improve the model's ability to find similar news stories across the remaining 17,300 news reports in our data set.[3] In our third and final round of model calibration, we manually coded an additional 200 news reports that the machine had already analyzed twice to determine the likelihood that they contained evidence of the conflict indicators in question. Once fed to the machine learning tool, this final round of training data was used to present to produce the trendlines shown throughout our Ukraine case study in the previous chapter.

---

[3] The technical appendix discusses our model calibration process in more detail.

This final round of training data allowed us to determine the accuracy of the machine learning model's results in the first two model runs. Specifically, we compared the predictions of the model during the first two calibration rounds with the manually coded values. In so doing, it helped us assess whether our machine was able to accurately assess evidence of the indicators in question, in comparison with the manual analysis.

Table 4.2 presents the statistical correlation between our manually coded indicators and the machine learning model's automatically coded estimates.[4] The first column compares the algorithm's results from the first model run with our eventual hand-coding of these news reports, after just 300 initial training data had been fed to the model. The second column uses

TABLE 4.2

Correlation Between Machine Learning Results and Hand-Coded Results

Indicator	Model Run #1	Model Run #2
Levels of violence	0.61	0.71
Russian involvement in violence	0.35	0.55
Ceasefire success	0.08	0.43
Ceasefire fail	0.33	0.30
Effective local governance	0.57	0.60
Ineffective local governance	0.50	0.51
Training data in model	300.00	700.00

NOTE: For each indicator, 200 news reports were coded as training data in the final (third) round of calibration. This table presents the statistical correlation between the human-coded values for each indicator from the third round of training data, and the machine-generated likelihood scores produced by the algorithm in the first two rounds of model calibration. The second run includes an additional 400 training data.

---

[4] The machine learning algorithm used in this research does not produce a binary *yes* or *no* result for each news report. Rather, it estimates the *probability* that any given news report contained evidence of the indicator in question. Therefore, it produces a simple score from zero to one that represents the likelihood that what the machine coded is similar to what was coded manually in the training data. The analysis in Table 4.2 presents the statistical correlation between these continuous likelihood scores from the first two rounds of model results and the binary manual coding across all 200 news reports for each indicator.

the algorithm's results from the second model run, after an additional 400 training data were provided to calibrate the machine learning model.

On their own, the correlation coefficients from both columns give a sense of the accuracy of the machine learning model relative to the manual analysis. In certain cases, such as the machine learning model's analysis of ceasefire successes, the model was only slightly more accurate than a random draw. In others, particularly the model's analysis of violence and effective local governance, the model's results were highly correlated with our eventual hand coding. Either way, these results clearly suggest that the machine learning tool was not as accurate as our human coding of the news reports.

To dig further into these results, we can assess our ability to train the model and improve its accuracy by comparing changes in these correlation coefficients across the two columns of the table. On the second model run, after feeding 400 additional training data to the machine learning model, the algorithm improved its correlation with our manual coding for five of the six conflict indicators. In one of these cases (ceasefire successes), the additional training data produced significant improvements in the model's accuracy. In other cases, the additional training data produced only marginal gains. And in the case of ceasefire failures, additional training data appear to have made the machine's model even less accurate. Ultimately, even after we trained the model through two rounds of additional training data, its results fail to perfectly align with our hand-coded results in all six indicators.

This hardly nullifies the usefulness of our machine learning approach. While producing the analysis presented in Chapter Three, we eventually concluded that this imprecision was not an obstacle to drawing useful implications from the trendlines the machine produced. Rather, we focused on the machine learning results simply as representative trends of the underlying operational environment.[5] To improve the accuracy of the machine

---

[5] We undertook a deliberate effort, detailed in the technical appendix, to test the sensitivity of our findings to alternative interpretations of the machine learning algorithm's likelihood scores. Throughout the report, the error bands presented around the trendlines for each indicator affirm that broad trends were robust to alternative likelihood thresholds.

learning model, an analyst could even leverage the significant time savings from a machine learning approach to manually code a larger proportion of underlying news reports, perhaps focusing on the reports the machine flagged as most likely to contain positive evidence of each indicator. In this way, a systems approach to human-machine collaboration can realize the full efficiency gains from machine learning while maximizing the accuracy of the ensuing results.

Either way, it is not at all certain that a human could improve on the accuracy of a machine when analyzing similarly large volumes of unstructured text. As shown in Table 4.1, we estimate that it would take 86 hours to manually review the 18,000 news reports used in our Ukraine demonstration. It is highly unlikely that a human analyst would apply coding criteria as consistently as a machine over this amount of time and across this volume of text.

Machine Learning Is Only as Good as the Underlying Data

Finally, it is worth emphasizing that machine learning–based assessments are only as good as the underlying data used to feed the machine learning tool. Machine learning tools provide a mechanism for rapidly extracting insights from data that are otherwise impossible to analyze systematically and objectively. But the value of the extracted insights depends on the underlying content and quality of the data. Although a lack of useful content can be frustrating, in that a machine is unlikely to reveal insights of any substantive value, this quality-of-data issue can become a major problem if the underlying data are biased in some way that is unknown to the analysts.

For one, this bias might manifest itself as underreporting of certain types of information, leading analysts to insights that simply reflect the way that the data were collected rather than actual events or conditions. We saw some evidence of this in our case study: Our news source seemed less likely to report ceasefire violations involving Ukrainian armed forces than those involving Russian or Russian-backed separatists.

Similarly, measurement error might result from differences in the language used to describe certain events compared with others. For example, we found that our machine learning tool was more confident in its ability to identify evidence of negative and adverse news events (such as failed cease-fires or ineffective governance) than in identifying positive news events (such as successful ceasefires or effective governance).[6] Although this could reflect the fact that negative news events are simply more newsworthy, it could also reflect that the machine learning tool found it easier to flag language describing clearly negative events and harder to parse language describing more nuanced, positive outcomes. Either way, this bias could incorrectly lead to the conclusion that ceasefire failures are more frequent than ceasefire successes or that ineffective governance is more frequent than effective governance.

Changes in the way that data are collected or documented can similarly introduce bias. We saw this type of bias clearly in our case study as the news source began publishing fewer details of violent incidents over time. Regardless of the cause (e.g., intentional editorial decision), the net result of this distributional shift in the data is to incorrectly imply that violence went down far more than other sources of data suggest to be true. Beyond just news reports, similar distributional shifts could affect machine learning's utility in analyzing other data more common to military headquarters. SITREPs, for example, could induce bias over time if the types of information captured in them were to change substantively year over year.

Although these potential biases (and others) do pose a challenge, they do not nullify the utility of machine learning for operational decisionmaking. Rather, they affirm the need to embrace a systems approach to machine learning that layers multiple sources of data alongside existing subject-matter expertise.

---

[6] The technical appendix to this report shows this in more detail across the various figures plotting the distribution of confidence scores that the machine learning tool produced across each of our indicators.

CHAPTER FIVE

Implications for the U.S. Army

Advances in machine learning have the potential to dramatically change the character of warfare by enhancing the speed, precision, and efficacy of decisionmaking across the national security enterprise. Leaders across the U.S. Department of Defense recognize this, and a multitude of efforts are underway to effectively integrate machine learning tools across the tactical, operational, strategic, and institutional levels of war. Throughout this report, we explored one application of machine learning, focused on enabling military decisionmaking at the operational level of competition and conflict. In this final chapter, we summarize three key findings and two recommendations for the Army that emerge from this research.

Research Findings

Human involvement—from analysts who possess detailed understanding of the context behind a given problem—is critical for deriving useful insights from currently available machine learning capabilities.

Human-machine collaboration is critical for using machine learning effectively in support of operational and institutional decisionmaking. Our case study demonstrated that using machine learning approaches to enable operational-level decisionmaking has great potential but also that relevant insights are possible only when machine learning tools are paired with human analysts who possess detailed understanding of the operational environment.

In our case study, most of the key insights gained through the use of machine learning required additional intervention from a human analyst. Naïve estimates using machine learning as a standalone tool, without added human context or insights, provided either an incomplete or misleading understanding of conditions on the ground in our analysis of the conflict in Ukraine. A more-sophisticated systems approach to machine learning, in which the machine learning tool works with a human analyst collaboratively and iteratively to analyze the operational environment, proved much more effective. Thus, it is likely that existing machine learning capabilities available to the U.S. Army will require humans to remain in the loop of the analytic process to unleash its full potential in operational-level decisionmaking.

By enabling dramatic efficiency gains in the performance of repetitive tasks, a human-machine systems approach can analyze massive data sets at a scale that would be impractical for human analysts alone, generating new insights about the operational environment that would previously have been unattainable.

Machine learning provides a capability for automating otherwise repetitive analytical tasks, leveraging modest amounts of training data from a human analyst, and extrapolating these insights against an even larger set of machine-coded data. In the pre-2022 Ukraine-focused analysis presented in this report, the machine learning approach was 40 percent more efficient than producing this analysis by hand, reducing the time spent on repetitive tasks by roughly six times. The analysts were more efficient, more rigorous, and better able to extract insights from previously untapped sources of data.

This suggests that Army leaders should prioritize machine learning as a solution to problems that require significant manual review of relevant data, such as intelligence analysis, talent management, or other fields in which analysts are trained to interpret large volumes of unstructured data to help make complex decisions.

Machine learning enables standardized, objective, and long-term analysis of a multitude of data streams already available in an operational-level headquarters.

Machine learning allows systematic and objective analysis of a variety of unstructured data that would typically require a human being to analyze by hand. Our analysis focused on unstructured text data, which have become ubiquitous as a result of the way that humans use technology. These data often offer the most nuanced and systematic understanding of diverse problem sets that military analysts and decisionmakers face, from recruiters reviewing candidates' personal statements, to intelligence analysts reviewing intercepted communications, to commanders reviewing SITREPs. Unstructured data exist in a variety of formats, and although the volume of unstructured data available to Army decisionmakers makes it impractical to analyze by hand, machine learning can enable analysts to rapidly and systematically extract insights from these data with relative ease.

The key implication of this finding is that machine learning can allow the U.S. Army to analyze new and previously untapped data sources. In many cases, these are the best sources of information for making decisions at the operational and institutional levels of war, but without machine learning, such data are left to be analyzed ad hoc and subjectively.

Recommendations

The Army should provide personnel at all echelons of command frequent exposure to machine learning, to build familiarity with how humans can leverage these capabilities as part of a military decisionmaking system.

Machine learning provides a powerful analytical capability for supporting military decisionmaking but one that requires human interaction at multiple points. As with early radar systems at the beginning of World War II, currently available machine learning algorithms are effective at flagging relevant trends based on observable data. But machine learning still requires the participation of leaders, operators, and analysts to design the algorithms, review the data and interpret the trends that emerge, and assess whether these trends are logically consistent with other available data.

The key implication here is that Army personnel at all echelons would benefit from exposure to practical examples of machine learning early, and often, throughout their careers. Even individuals who are unlikely to implement machine learning algorithms themselves remain likely to make the types of decisions that could be enabled by machine learning. Without exposure to the strengths and limitations of machine learning in the context of actual defense problems, soldiers and civilians alike may fail to explore chances for innovation or fail to tackle pressing problems, hoping that some machine may eventually do the hard work for them.

The Army should build diverse machine learning teams to unlock the full potential of this capability, integrating ORSAs with operators, analysts, and commanders to help interpret the implications of machine learning-based analysis.

In our analysis of the conflict in Ukraine prior to Russia's 2022 full-scale invasion, we proposed a notional machine learning team at an operational-level headquarters, comprising intelligence analysts, operators, and planners who deeply understood their operating environment. Data science expertise was necessary to implement the algorithm itself, but the bulk of the work involved individuals with knowledge of the Ukrainian operating environment manually coding news reports to identify evidence of violence, ceasefires, and local governance to train the machine learning tool.

To that end, the best machine learning teams in the Army will include personnel with diverse experience in an operational environment and the ability to understand the relevance of the information that goes into and comes out of the machine learning tool. The best radar operators are not physicists or mechanical engineers skilled in the science of sound-wave propagation but are those with experience interpreting the feedback provided by radar systems against real-world conditions. The same applies to machine learning.

As a result, Army commands and staff sections looking to stand up machine learning teams should build these teams from a variety of personnel with diverse backgrounds not just those able to code their own machine learning algorithms from scratch. A careful balance of technical expertise and operational context is required to unleash machine learning's full potential.

Training a Machine to Read the News

This appendix provides a technical overview of our machine learning analysis of news reports about the conflict in eastern Ukraine prior to Russia's 2022 invasion. It seeks to provide both a detailed accounting of how we produced the results presented in Chapter Three, as well as a deeper contextual understanding of the strengths and limitations of using a machine learning approach grounded in a more detailed yet accessible discussion of a machine learning approach in action. It narrates the deliberately simple and transparent machine learning process used to produce our Ukraine case study presented in Chapter Three—meant to resemble the type of basic machine learning analysis that the staff of an operational-level headquarters could conduct with little to no formal data science training, This appendix begins with a contextual discussion of machine learning, followed by a discussion of the data source used in the analysis, the process used to extract relevant information from that data source, coding criteria used to develop training data for each indicator, and finally the simplified calibration process used to improve the machine learning algorithm's performance.

Machine Learning in Context

To understand how a machine can be trained to "read" in this context, it is helpful to think first about how humans communicate themselves. Humans learn to communicate, in part, by learning to organize specific words with specific meanings into larger sentences. Understanding a sentence boils down to two things—the words used in the sentence and how the words are organized relative to each other.[1] Machines learn to read unstructured text in much the same way—by identifying how frequently certain words are used in proximity to others. But unlike humans, machines do not know the meaning of the individual words they read. As a result, humans must "train" the machine to associate certain words and patterns among them with some larger meaning. This is the learning aspect of machine learning.

Selecting the Right Data

News reports can, conceptually, be a useful tool for machine learning analysis. They provide a consistent historical record of events over time, using descriptive words and phrases that are useful fodder for training a machine. And although a human can read a sample of news reports on their own, it is often too burdensome for a human to read *every* news report about a topic over some period. That is where machine learning can help.

To find an appropriate historical record of the conflict in Ukraine for a machine learning analysis, we focused on one specific authoritative source—UKRINFORM. Using Nexis-Uni, we downloaded every English-language news report from UKRINFORM since 2014 that mentioned "Donetsk," "Luhansk," or "eastern Ukraine."[2] Nexis-Uni is a subscription-based service that aggregates news reporting from various sources in most countries across the globe, across multiple languages. In practice, Army users looking to analyze news reports could rely on existing U.S. government databases for open-source research (such as the Open Source Enterprise).

This produced a rich library of 17,995 separate news articles describing the conflict. Assembling this database required both manual effort (to download the reports in small batches) and some technical effort (to translate

---

[1] As an example, the phrase "I ran out to the store to buy milk" means something substantively different from "the store ran out of milk," despite sharing five of the same words. Our understanding of the difference between these phrases is driven by two factors—the added word used in the first sentence ("buy") and the varied ordering of the words between the two sentences ("ran out to the store" versus "the store ran out").

[2] We tested the sensitivity of the results of this search to using alternative terms (e.g., "Donbas") and found only marginal differences in the resulting set of news reports describing the conflict.

each news report into a machine-readable format), both of which could be conducted by minimally trained Army staff under the right circumstances.[3]

Relying on a single source of news reports can be beneficial for machine learning based on news, in that the language used in any one data point is likely to share a common lexicon and editing style with the full body of other textual data in the larger data set. Similarly, a single source of news is unlikely to include repetitive or duplicative reporting of identical events. However, reliance on one news source could bias a machine learning analysis if that source failed to report on underlying historical events consistently over time.

The selection of UKRINFORM should mitigate this risk, given its status as a national news service with international reach. Similarly, editorial bias is a consideration; for instance, UKRINFORM might overinflate the severity of Russian and separatist violence and underinflate Ukrainian failures. However, as long as this bias is generally consistent, it should not affect our results when comparing trends over time.

Extracting Relevant Information

We next identified the portions of UKRINFORM reports that were most likely to contain relevant information for our analysis about the conflict. To do so, we extracted each sentence from the broader news report that mentioned a relevant keyword, as well as the sentence before it and the sentence after it to provide additional context.[4] Together, these three sentence

---

[3] Downloads from Lexis-Nexis are limited to a specific batch size; downloading = the full time-series of data thus required significant time. Once the data had been downloaded, extracting the metadata and content from each news report required additional effort (using statistical analysis software). In practice, the Army could develop and proliferate basic tools to consolidate news reports from relevant databases (either Lexis Nexis, Open Source Enterprise, or other sources) into a usable format for machine learning.

[4] Although we used statistical analysis software to do this task, this could, in practice, just as easily be done by an Army soldier or civilian minimally trained in software like Microsoft Excel.

extracts represent relevant news mentions of specific events and form the basis of our machine learning analysis.

To choose the right keywords, we first read a sample of articles to understand how the source of the news, UKRINFORM, was likely to talk about the types of events we were looking to measure—violence, ceasefire effectiveness, and local governance. When discussing violence initiated by separatist forces, UKRINFORM often labeled these incidents as "attacks" or mentioned "shelling" or "bombardment" by separatist forces. When discussing progress toward ceasefires, UKRINFORM often mentioned the "withdrawal" of forces from the front lines, or "peaceful" conditions on the ground following a "ceasefire."[5] When discussing the quality of local governance in the separatist regions, we noted that UKRINFORM articles were likely to mention damage to certain types of "infrastructure," including "hospitals," "schools," "banks," and "power plants." We also noticed frequent discussion about access to state "pensions" for Ukrainian civilians inside the Donbas region and Russian "humanitarian" convoys to provide economic aid (and, often, military support).[6]

There were certainly other keywords that we could have used to extract relevant news mentions from the larger data set of news reports. But we discovered that, more often than not, additional words that we did not select appeared right alongside the keywords we did use to extract the right data.[7]

---

[5] After we identified keywords relevant to both levels of violence and ceasefire effectiveness, we determined that there was significant overlap between the sentences describing each type of event. As a result, we combined the two sets of keywords into one list. The exact list of keywords used to extract sentences for these indicators was "attack," "shell," "bomb," "explosion," "shoot," "peace," "ceasefire," "violence," and "withdrawal." Note that some words were shortened (e.g., "shell") to help capture various different conjunctions of the word (e.g., "shelling," "shelled," "shell").

[6] In all, we used 15 separate keywords to extract 12,149 sentences of interest, drawn from 4,956 articles, or 28 percent of the 18,000 articles in our sample. The exact list of keywords used to extract sentences for this indicator was "hospital," "pension," "bank," "transport," "infrastructure," "humanitarian," "medical," "school," "communication," "phone," "power," "water," "transformer," "electric," and "plant."

[7] To measure this, we did basic keyword searches for other relevant terms for each indicator to assess whether any new term should be included in our searches. In some cases, this helped identify additional keywords that extracted truly relevant data worth including in our sample. But when additional keywords added relatively few new sen-

Moreover, the purpose of this step was not to use the keywords themselves to measure trends. Rather, it was to give the *machine* the best possible set of data to measure trends over time.[8]

Training the Machine Learning Tool and Calibrating the Algorithm

With the relevant news mentions extracted from our broader library of news reports, we next turned to training the machine learning tool to identify the specific types of events we care the most about: violence, ceasefire effectiveness, and local governance.

Levels of Violence and Russian Involvement

To understand how levels of separatist violence changed over time since the start of the conflict, we trained the machine learning tool to identify specific violent incidents in UKRINFORM news reports.[9] To do so, we first had to manually code a set of training data, drawn from a random subset of the

---

tences of little relevance, we were satisfied. After all, the purpose of a machine learning approach is to avoid the need for a human analyst to assess every piece of data manually.

[8] For example, some portions of UKRINFORM news reports that mention "Luhansk" or "Donetsk" or "eastern Ukraine" were relatively benign articles about sports, culture, business, and local events. By explicitly removing these articles from our sample, we allowed the machine to focus solely on the words, phrasings, and style of language used to specifically discuss the aspects of the conflict we care the most about. This should reduce statistical noise in our analysis, to the extent that articles about benign events may have linguistic similarities with events describing the conflict. Imagine, for example, conflating an article that narrates a football team's defeat in a game with another article that discusses the defeat of separatist forces in battle.

[9] Several alternative methods are available for measuring levels of violence in conflict. Where the U.S. military has historically had its own forces actively engaged in combat (e.g., in Iraq and Afghanistan), it has collected data on significant activities that measure enemy-initiated attacks, among other events. A number of publicly available academic data sets also track attacks, such as ACLED and the Janes Terrorism and Insurgency Database (Janes Information Services, 2020). Each of these data sets relies on largely manual aggregation of human observations, whether from news reports or physical observation.

larger set of relevant news reports, so that the machine learning tool could flag similar events in the remainder of the news reports on its own.[10] Within this random sample of training data, we looked for discussion of a specific attack, bombardment, or skirmish initiated by separatist forces (or Russian forces) against Ukrainian armed forces or civilians.[11] While reviewing the data, we noted that these news reports often provided clear evidence of deliberate Russian support to violence against Ukrainian armed forces and civilians, including through logistical support provided to the separatists, artillery support from across the border, and direct provision of military equipment. As a result, we chose to train the machine learning tool to identify evidence of Russian involvement as a separate, distinct indicator.[12]

After manually coding the training data, we fed them into our machine, running a deliberately simple off-the-shelf algorithm capable of basic natural language processing tasks that compared the words used in our training data with the words used across all the potentially relevant sentences drawn from the 18,000 news reports.[13] The result from this analysis was not an authoritative *determination* of how many attacks separatist forces committed or whether Russia was involved in that violence. Rather, the result

---

[10] Of the nearly 8,000 sentences possibly mentioning attacks or violence, we at first randomly selected 300 of these sentences, flagging those that provided concrete evidence of attacks by separatist forces. Of these 300, more than one-half (154) identified examples of separatist violence against Ukrainian armed forces, civilians, or critical infrastructure. This is perhaps unsurprising, given the keywords we used to extract these sentences from the broader 18,000 news reports in our sample.

[11] It was not always possible to determine who initiated the violence. As a result, we included any discussion of violence involving the separatists that did not directly state that the Ukrainian Armed Forces had initiated the violent incident.

[12] Of the 300 cases in our initial round of training data, we identified only 22 instances of direct Russian support to violence.

[13] We use a kernel support vector machine (ksvm) algorithm, publicly available via the Kernlab package in the R statistical software suite, to replicate the types of baseline methods that could be reliably employed by the staff of an operational-level military headquarters with little to no formal data science expertise. See Alexandros Karatzoglou, Alex Smola, Kurt Hornik, National ICT Austrial, Michael A. Maniscalco, and Choon Hui Teo, "Kernel-Based Machine Learning Lab," Comprehensive R Archive Network, undated.

was the machine learning model's estimate of the *probability* that any given news report contained evidence of such events.

How confident was our machine that it was accurately reading the news? For each news report, the algorithm produced a simple score from zero to one that represented the likelihood that what the machine coded was similar to what we had coded manually in the training data as evidence of an attack or of Russian involvement in violence. Figure A.1 plots the distribution of likelihood scores for both indicators by percentile, with violence in blue and Russian involvement in green. In this figure, the darkest blue line represents the machine learning model's relative certainty that any given news report provided evidence of a violent attack after our initial batch of training data. The darkest green line reports the same for evidence of Russian involvement. The x-axis represents the percentile of each score in the

FIGURE A.1

Machine-Generated Likelihood Scores for Levels of Violence and Russian Involvement

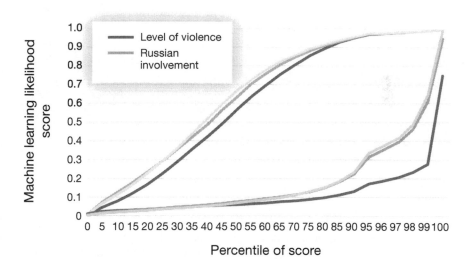

SOURCE: Authors' calculations based on news reporting from UKRINFORM news reports accessed through the Nexis-Uni academic research tool.

NOTE: For each indicator (in blue and green), the darkest color represents results from the first run of the machine learning algorithm, and the lighter colors represent each of the next two runs following calibration.

broader distribution of likelihood scores for each indicator.[14] Plotted in this manner, Figure A.1 presents a hybrid measure of the proportion of news mentions most likely to contain evidence of each indicator.

These results suggest two things. First, the machine learning tool was more confident in its ability to identify widespread evidence of separatist violence across more news reports than to identify Russian support to that violence at similar volumes. This is demonstrated by the higher proportion of news mentions (along the x-axis) with higher machine-generated likelihood scores for that indicator (along the y-axis). This is unsurprising, in that discussion of Russian involvement in violence is more nuanced than discussion of the violence itself. Additionally, this analysis suggests that the machine learning tool was confident only that a tiny fraction of news reports discussed Russian support to the separatists. This is more surprising, given that accusations of Russian support to the separatists are a defining feature of the conflict.

Thankfully, we can help the machine learning tool improve its ability to diagnose these events by providing it more training data.[15] In the context of this figure, the purpose of calibrating the algorithm is to ensure that news mentions not showing evidence of each indicator receive lower likelihood scores and that news mentions showing evidence of each indicator receive higher likelihood scores.[16] So in a second run of our algorithm, we fed an additional 400 rows of training data to the machine learning model for both indicators. As shown in Figure A.1, this produced only modest gains in the

---

[14] For example, the median likelihood score (50th percentile) from the first model run for levels of violence was 0.568.

[15] When choosing additional training data, we employed a deliberately simplified and transparent approach to model calibration and validation that randomly selected additional news reports in two categories according to the likelihood scores in the first model for both indicators. For one-half of the additional training data, we manually coded news reports that had high likelihood scores to minimize the likelihood of false positives. For the other half, we selected news reports that had medium likelihood scores to minimize the likelihood of including false negatives above a certain threshold.

[16] In a perfect scenario, this would produce an S-shaped curve of likelihood scores, with some proportion of news mentions demonstrating very low likelihood scores and a larger proportion receiving higher scores. The inflection point between these scores would resemble the natural incidence rate of each indicator in the underlying data.

ability of the machine learning model to diagnose violent incidents in a higher proportion of news mentions but did produce significant improvements in its ability to diagnose Russian involvement in the conflict. In a third attempt, we fed an additional 200 rows of training data into model but produced only marginal gains in likelihood scores across the two indicators. These declining marginal gains in confidence suggested that additional training data beyond a third round would not likely improve the model's calibration.

So, what do we do with these likelihood scores? After all, the machine learning model, at this point, has only expressed its confidence that any one news report actually represents the type of violence that we are hoping to measure. One option would be to identify the news reports below a certain likelihood score (likely positive evidence) and above a certain likelihood score (likely negative evidence), for which we could outright accept the machine learning model's results within a certain margin of error. Then, we could manually code each news report left in the middle.[17] However, in our case, this left a significant amount of data to be manually coded, potentially reducing the utility of using a machine learning approach in the first place.[18]

An alternative approach, the one we chose to employ for these indicators and throughout the rest of our analysis, is to simply set a likelihood score above which we are reasonably confident in the machine learning model's results. We then assess whether overall trends change when we adjust that threshold up or down. This approach is ideal, in that it does not require fur-

---

[17] In practice, this approach requires arbitrarily selecting thresholds from the data based on the likelihood scores produced by the machine learning algorithm and then assessing the rate of false positives or negatives above and below these thresholds. We attempted this approach, allowing a 5-percent margin of error. That is, we set a minimum likelihood threshold below which no more than 5 percent of a random sample of the sentences flagged by the machine learning algorithm as low-probability events offered actual evidence of violence (false negatives). Similarly, we set a maximum likelihood threshold above which no greater than 5 percent of a random sample of the sentences flagged by the algorithm as high probability events offered no evidence of violence (false positives).

[18] Across all of the indicators in this case study (violence, ceasefires, and governance), we determined that this approach would require manually coding between 9 and 30 percent of all the news reports in our data set, arguably defeating the purpose of a machine learning approach. This option could be more applicable in other analyses where the algorithms produce greater differentiation between positive and negative results.

ther manual coding of data beyond the training data. But it is only appropriate when the trendlines the machine learning model produces are robust to different selections of the likelihood score above which we choose to accept the machine learning model's results. And it does limit our ability to draw insights from marginal differences in the magnitude of news mentions *across* indicators.

We selected appropriate thresholds for likelihood scores by identifying the minimum threshold below which a random sample of machine-coded news reports did not contain more than 5-percent false negative results, then tested the sensitivity of our results to significantly more-restrictive specifications. We found that this approach produced trendlines that were both informative and robust to different specifications of the likelihood scores. Therefore, we report levels of violence in the body of this report that are based on the news mentions for which the algorithm's likelihood score exceeded the 25th percentile of likelihood scores. Alternative specifications are shown in the body for news mentions above the 15th percentile and 50th percentile. The differences are minor between the three total specifications. For Russian involvement in violence, we report results based on the news mentions for which the algorithm's likelihood score exceeded the 90th percentile. Alternative specifications are shown for news mentions above the 80th percentile and above the 95th percentile. The trends for both indicators were robust to these alternative specifications.[19]

In the body of this report, we also compare our machine learning results to the levels of violence tracked by other data sources, including the Janes Terrorism and Insurgency Database,[20] and ACLED.[21] We found that the

---

[19] For both indicators, the thresholds we selected fell slightly below the rate at which each indicator occurred in the original random sample of training data. Of the original 300 rows of training data, we coded 154 news mentions as providing evidence of violent incidents (51.3 percent) and 22 news mentions as providing evidence of Russian involvement in violence (7.3 percent). Although these sampling rates are not precise measures of the underlying incidence rate in the entire data set, they do provide a helpful robustness check for us to assess the sensitivity of our findings.

[20] Janes Information Services, 2020.

[21] Clionadh Raleigh, Andrew Linke, Håvard Hegre, and Joakim Karlsen, "Introducing ACLED-Armed Conflict Location and Event Data," *Journal of Peace Research*, Vol. 47, No. 5, September 2010.

rapid decline in levels of violence reported in our machine learning analysis of news significantly underreported the levels of violence in recent years. However, each of these data sets confirms our underlying finding that violence has declined in recent years.

Table A.1 compares the volume of violent incidents that these alternative sources of data reported per year with the numbers from our analysis. Our machine learning analysis appears to provide better fidelity and variation on levels of violence in the first two years of the conflict (2014–2015) than the alternative sources do. Although Janes suggests that the level of violence in these years was considerably lower, this may be a product of their data generation strategy, which focused on measuring distinct attacks and could struggle to disaggregate incidents of violence that occurred as part of larger confrontations in the early, more active years of the conflict.

TABLE A.1

Comparison of Levels of Violence Across Data Sets

Year	Machine Learning	Janes	ACLED
2014	1,525	426	
2015	2,179	1,153	
2016	1,078	4,466	
2017	810	3,741	
2018	106	4,435	13,131
2019	90	3,422	14,663
2020 (Q1–Q2)	39	1,505	6,510

SOURCE: Authors' calculations based on news reporting from UKRINFORM, made available via Nexis-Uni; Janes Information Services, 2020; Raleigh et al., 2010.

NOTE: Data from ACLED are not available prior to 2018. For Janes data, we report all nonstate armed group attacks occurring inside Ukraine over this time period. For ACLED, we report both "battles" and "explosions/remote violence" based on their coding criteria.

Assessing Ceasefire Effectiveness

To assess ceasefire effectiveness, we followed the same general approach as we did for understanding levels of violence, as described in the previous section. The focus here is on two separate measures of ceasefire effectiveness: whether a news report provides concrete evidence of successful adherence to the terms of a ceasefire and whether the report provides concrete evidence of failure to adhere to the terms of a ceasefire. We focus specifically on evidence that separatist forces (and Russia), rather than Ukrainian armed forces, are upholding or violating terms of a ceasefire.[22]

Using a machine learning approach to assess ceasefire effectiveness offers an interesting comparison with our analysis of levels of violence. For one, these two indicators should be highly correlated, to the extent that such attacks often constitute a violation of a ceasefire. Yet, beyond measuring discrete events, our machine learning approach can also pick up on discussion of things that *did not happen*. For example, our approach can identify instances of news reports discussing the failure of separatist forces to withdraw from the front line as evidence of a failed ceasefire, while also flagging news reports about relative calm after a ceasefire as evidence of success.

When coding training data as evidence of ceasefire success, we looked for evidence of the withdrawal of separatist forces from the front lines, discussion of reduction in violence after a ceasefire takes effect, references to stability or peace during a ceasefire, and implementation of certain ceasefire terms (such as prisoner swaps). When coding training data as evidence of ceasefire failure, we looked for allegations or concrete evidence of attacks that violate a ceasefire, use of prohibited weapons when specifically highlighted as a ceasefire violation, refusal to execute certain terms of the ceasefire (such as prisoner swaps), and separatist efforts to deny OSCE conflict observers access to conflict-affected zones.

In three successive rounds of model calibration, we manually coded 900 sentences drawn randomly from news reports discussing ceasefires as train-

---

[22] We do so for two reasons. First, UKRINFORM provides an inconsistent appraisal of where ceasefires have been violated by Ukrainian armed forces. Second, from a U.S. perspective, this analysis is conceptually focused on understanding the adversary's behavior rather than the partner's behavior.

ing data.[23] Figure A.2 plots the distribution of likelihood scores for both indicators, with successful ceasefires in blue and failed ceasefires in green. The x-axis represents the percentile of each score in the broader distribution of likelihood scores for each indicator.[24] Plotted in this manner, Figure A.2 presents a hybrid measure of the proportion of news mentions most likely to contain evidence of each indicator.

FIGURE A.2

Machine-Generated Likelihood Scores for Ceasefire Effectiveness

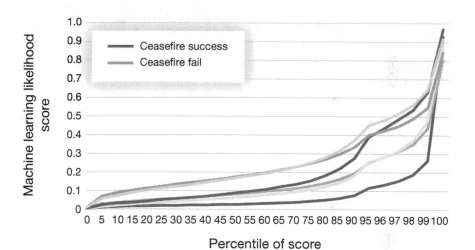

SOURCE: Authors' calculations based on news reporting from UKRINFORM news reports accessed through the Nexis-Uni academic research tool.

NOTE: For each indicator (in blue and green), the darkest color represents results from the first run of the machine learning algorithm, and the lighter colors represent each of the next two runs following calibration.

---

[23] We used 300 rows of training data in the first model run, an additional 400 in the second, and an additional 200 in the third. The first set of training data was randomly drawn from the entire sample of news mentions. For both additional runs of the algorithm, we employed a different approach for selecting the training data—selecting random samples from both high-likelihood news mentions (to minimize the likelihood of false positives) and medium-likelihood news mentions (to minimize the likelihood of false negatives).

[24] For example, the median likelihood score (50th percentile) from the first model run for successful ceasefires was 0.029.

As before, the first run of the algorithm is plotted in darker colors, and the successive runs in lighter shades. The algorithm appears much more confident in its ability to identify evidence of failed ceasefires than ceasefire successes. Also, as before, additional training data improved the confidence of the algorithm up to a certain point. Notably, however, the final (third) set of training data actually reduced the algorithm's confidence in its analysis of failed ceasefires. This affirms the declining marginal utility of additional training data in some cases and the importance of conducting diagnostic analyses when using machine learning to properly calibrate each model.

To translate these likelihood scores into the trendlines shown in the body of the report, we set maximum thresholds above which we were reasonably confident in the machine learning model's results and assessed whether overall trends changed if we adjusted that threshold up or down.

For ceasefire successes, our results are based on the news mentions for which the algorithm's likelihood score exceeded the 90th percentile. Alternative specifications are shown for news mentions above the 80th percentile and above the 95th percentile. For ceasefire failures, our results are based on the news mentions for which the algorithm's likelihood score exceeded the 70th percentile. Alternative specifications are shown for news mentions above the 60th percentile and above the 80th percentile. The trends for both indicators were robust to these alternative specifications.[25]

Assessing the Quality of Local Governance in the Donbas Region

We concluded our analysis by employing the same machine learning approach to search for evidence in news reports that the separatist governments have provided effective or ineffective local governance.

---

[25] For both indicators, the thresholds we selected fell slightly below the rate at which each indicator occurred in the original random sample of training data. Of the original 300 rows of training data, we coded only 13 news mentions as providing evidence of successful ceasefires (4.3 percent) and 37 news mentions as providing evidence of failed ceasefires (12.3 percent). Although these sampling rates are not precise measures of the underlying incidence rate in the entire data set, they do provide a helpful robustness check for us to assess the sensitivity of our findings.

We began by manually coding a random subset of news reports for evidence of these two indicators to build training data to provide to the machine learning tool. When coding training data as evidence of effective governance, we looked for evidence of successful public-service provision inside separatist-held territory, including public hospitals, banks, transportation services, communications infrastructure, and other government-furnished or government-regulated services. Because of the limitations of how these services were described in UKRINFORM reporting, we remained agnostic as to whether these services were provided by the separatist regimes or by Ukrainian government entities with reach into the separatist-controlled territories.[26] We also looked for evidence of Russian humanitarian or economic support to the rebel-held regions. Additionally, we noted instances of news reports discussing economic growth or the restoration of critical infrastructure inside separatist-held regions.

When coding training data as evidence of ineffective governance, we looked for evidence of inadequate public service provision inside separatist-held territory, including the lack or insufficiency of public hospitals, banks, transportation services, communications infrastructure, and other government-furnished or government-regulated services. We also looked for evidence of deprivation caused by conflict and of the destruction of critical infrastructure caused by separatist-initiated violence. Anecdotally, destruction of critical infrastructure comprised a significant percentage of the instances of ineffective governance in our data set, primarily in the early years of the conflict.

We then calibrated our machine learning algorithm across three successive rounds, each time feeding it additional training data to improve its ability to diagnose effective and ineffective local governance.[27] Figure A.3 plots the distribution of likelihood scores for both indicators. The x-axis

---

[26] The machine proved unable to differentiate between public services provided by rebel-controlled entities and those provided by Ukrainian entities based solely on the content of individual news articles.

[27] We used 300 rows of training data in the first model run, an additional 400 in the second, and an additional 200 in the third. We drew first set of training data randomly from the entire sample of news mentions. For both additional runs of the algorithm, we employed a different approach to selecting the training data—selecting random samples from both high-likelihood news mentions (to minimize the likelihood of false

FIGURE A.3

Machine-Generated Likelihood Scores for Quality of Local Governance

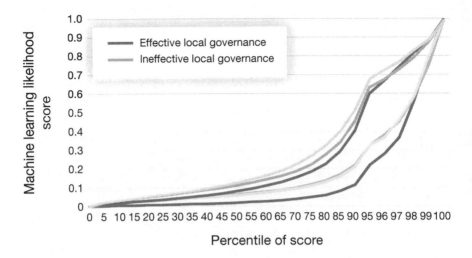

SOURCE: Authors' calculations based on news reporting from UKRINFORM, made available via Nexis-Uni.

NOTE: For each indicator (in blue and green), the darkest color represents results from the first run of the machine learning algorithm, and the lighter colors represent each of the next two runs following calibration.

represents the percentile of each score in the broader distribution of likelihood scores for each indicator.[28] Plotted in this manner, Figure A.3 presents a hybrid measure of the proportion of news mentions most likely to contain evidence of each indicator.

To translate these likelihood scores into the trendlines shown in the body of the report, we set maximum thresholds above which we were reasonably confident in the machine learning model's results and assessed whether overall trends would change if we adjusted that threshold up or down.

---

positives) and medium-likelihood news mentions (to minimize the likelihood of false negatives).

[28] For example, the median likelihood score (50th percentile) from the first model run for successful ceasefires was 0.018.

For effective local governance, our results are based on the news mentions for which algorithm's likelihood score exceeded the 90th percentile. Alternative specifications are shown for news mentions above the 80th percentile and above the 95th percentile. For ineffective local governance, our results were based on the news mentions for which algorithm's likelihood score exceeded the 70th percentile or higher. Alternative specifications are shown for news mentions above the 60th percentile and above the 80th percentile. The trends for both indicators were robust to these alternative specifications.[29]

---

[29] For both indicators, the thresholds we selected fell slightly below the rate at which each indicator occurred in the original random sample of training data. Of the original 300 rows of training data, we coded only 17 news mentions as providing evidence of effective local governance (5.6 percent) and 49 news mentions as providing evidence of ineffective local governance (16.3 percent). Although these sampling rates are not precise measures of the underlying incidence rate in the entire data set, they do provide a helpful robustness check for us to assess the sensitivity of our findings.

Abbreviations

ACLED	Armed Conflict Location Event Dataset
DPR	Donetsk People's Republic
G2	intelligence directorate
G3	operations directorate
LPR	Luhansk People's Republic
ORSA	operations research systems analyst
OSCE	Organization for Security and Co-operation in Europe
POLAD	political advisor
SITREP	Situation Report
SMM	Special Monitoring Mission
UKRINFORM	Ukrainian National News Agency
USEUCOM	U.S. European Command

References

Assessing Revolutionary and Insurgent Strategies Project, *"Little Green Men":*
A Primer on Modern Russian Unconventional Warfare, Ukraine 2013–2014,
U.S. Special Operations Command, 2015. As of July 1, 2021:
https://www.soc.mil/ARIS/books/arisbooks.html

Atherton, Kelsey, "Targeting the Future of the DoD's Controversial Project
Maven Initiative," C4ISRNET, July 27, 2018. As of September 13, 2018:
https://www.c4isrnet.com/it-networks/2018/07/27/
targeting-the-future-of-the-dods-controversial-project-maven-initiative/

Berruti, Federico, Pieter Nel, and Rob Whiteman, "An Executive Primer on
Artificial General Intelligence," McKinsey & Company, April 29, 2020. As of
September 8, 2020:
https://www.mckinsey.com/business-functions/operations/our-insights/
an-executive-primer-on-artificial-general-intelligence

Callandar, Bruce D., "The Ground Observer Corps," *Air Force Magazine*,
February 1, 2006. September 8, 2020:
https://www.airforcemag.com/article/0206goc/

Clymer, Kenton, "The Ground Observer Corps: Public Relations and the Cold
War in the 1950s," *Journal of Cold War Studies*, Vol.15, No. 1, Winter 2013.

Dear, Keith, "A Very British AI Revolution in Intelligence Is Needed," War on
the Rocks, October 19, 2018.

Egel, Daniel, Ryan Andrew Brown, Linda Robinson, Mary Kate Adgie, Jasmin
Léveillé, and Luke J. Matthews, *Leveraging Machine Learning for Operation
Assessment*, RAND Corporation, RR-4196-A, 2022. As of July 31, 2023:
https://www.rand.org/pubs/research_reports/RR4196.html

Fischer, Sabine, *The Donbas Conflict: Opposing Interests and Narratives,
Difficult Peace Process*, Berlin: German Institute for International and Security
Affairs, April 2019. As of September 8, 2020:
https://www.swp-berlin.org/10.18449/2019RP05/#en-d16368e1017

Florea, Adrian, "Rebel Governance in De Facto States," *European Journal of
International Relations*, Vol. 26, No. 4, December 2020.

Gibbons-Neff, Thomas, "Three-Day-Old Ceasefire in Ukraine Broken as
Fighting Resumes in some Areas," *Washington Post*, September 3, 2015.

Hoffman, Frank, "The Hypocrisy of the Techno-Moralists in the Coming Age
of Autonomy," War on the Rocks, March 6, 2019.

Imperial War Museums, "Support from the Ground in the Battle of Britain," webpage, undated. As of September 8, 2020:
https://www.iwm.org.uk/history/
support-from-the-ground-in-the-battle-of-britain

Janes Information Services, "Terrorism and Insurgency Database," 2020.

Karatzoglou, Alexandros, Alex Smola, Kurt Hornik, National ICT Austrial, Michael A. Maniscalco, and Choon Hui Teo, "Kernel-Based Machine Learning Lab," Comprehensive R Archive Network, undated.

Karlin, Mara, *The Implications of Artificial Intelligence for National Security Strategy,* Brookings Institution, November 1, 2018.

Kofman, Michael, Katya Migacheva, Brian Nichiporuk, Andrew Radin, Olesya Tkacheva, and Jenny Oberholtzer, *Lessons from Russia's Operations in Crimea and Eastern Ukraine*, RAND Corporation, RR-1498-A, 2017. As of July 1, 2021:
https://www.rand.org/pubs/research_reports/RR1498.html

Lim, Nelson, Bruce R. Orvis, and Kimberly Curry Hall, *Leveraging Big Data Analytics to Improve Military Recruiting*, RAND Corporation, RR-2621-OSD, 2019. As of July 1, 2021:
https://www.rand.org/pubs/research_reports/RR2621.html

"Luhansk Energy Association to De-Energize Popasnianskyi District Water Supply Channel Supplying with Water 75,000 Residents of Luhansk Region on February 21 Due to UAH 24.3 Million Debts," Ukrainian News Agency, February 17, 2020.

McKendrick, Kathleen, "The Application of Artificial Intelligence in Operations Planning," paper presented at the 11th NATO Operations Research and Analysis (OR&A) Conference, October 9, 2017.

"New Year Ceasefire Enters into Force in Donbass," TASS Russian News Agency, December 28, 2018. As of September 2018:
https://tass.com/world/1038447

Niland, Paul, "Making Sense of Minsk: Decentralization, Special Status, and Federalism," Atlantic Council, January 27, 2016.

Organization for Security and Co-operation in Europe Special Monitoring Mission to Ukraine, "2019 Trends and Observations," webpage, 2019. As of September 8, 2020:
https://www.osce.org/files/f/documents/1/e/444745.pdf

Organization for Security and Co-operation in Europe Special Monitoring Mission to Ukraine, "Trends and Observations: Jan–Mar 2020," webpage, 2020a. As of September 8, 2020:
https://www.osce.org/files/f/documents/0/d/450175.pdf

Organization for Security and Co-operation in Europe Special Monitoring Mission to Ukraine, "Trends and Observations: Apr–Jun 2020," webpage, 2020b. As of September 8, 2020:
https://www.osce.org/files/f/documents/e/d/457987.pdf

OSCE SMM Ukraine — *See* Organization for Security and Co-operation in Europe Special Monitoring Mission to Ukraine.

Paul, Christopher, Colin P. Clarke, Bonnie L. Triezenberg, David Manheim, and Bradley Wilson, *Improving C2 and Situational Awareness for Operations in and Through the Information Environment*, RAND Corporation, RR-2489-OSD, 2018. As of July 1, 2021:
https://www.rand.org/pubs/research_reports/RR2489.html

Pellerin, Cheryl, "Project Maven to Deploy Computer Algorithms to War Zone by Year's End," press release, U.S. Department of Defense, July 21, 2017.

Raleigh, Clionadh, Andrew Linke, Håvard Hegre, and Joakim Karlsen, "Introducing ACLED-Armed Conflict Location and Event Data," *Journal of Peace Research*, Vol. 47, No. 5, September 2010.

Robinson, Eric, Daniel Egel, Patrick Johnston, Sean Mann, Alex Rothenberg, and David Stebbins, *When the Islamic State Comes to Town: The Economic Impact of Islamic State Governance in Iraq and Syria*, RAND Corporation, RR-1970-RC, 2017. As of July 1, 2021:
https://www.rand.org/pubs/research_reports/RR1970.html

Robinson, Linda, Daniel Egel, and Ryan Andrew Brown, *Measuring the Effectiveness of Special Operations*, RAND Corporation, RR-2504-A, 2019. As of July 1, 2021:
https://www.rand.org/pubs/research_reports/RR2504.html

Ross, Casey, and Ike Swetlitz, "IBM Pitched Its Watson Supercomputer as a Revolution in Cancer Care. It's Nowhere Close," STAT, September 5, 2017. As of September 8, 2020:
https://www.statnews.com/2017/09/05/watson-ibm-cancer/

"Russia Sends 98th Humanitarian Convoy to Donbas, Ukraine's Foreign Ministry Protests," 112 Ukraine, August 28, 2020. As of September 8, 2020:
https://112.international/conflict-in-eastern-ukraine/russia-sends-98th-humanitarian-convoy-to-donbas-ukraines-foreign-ministry-protests-54243.html

Sayler, Kelley M., *Artificial Intelligence and National Security*, Congressional Research Service, R45178, August 26, 2020.

Scaparrotti, Curtis, "USEUCOM 2019 Posture Statement," testimony before the Senate Armed Services Committee, March 5, 2019. As of September 8, 2020:
https://www.eucom.mil/article/39546/useucom-2019-posture-statement

Schuety, Clayton, and Lucas Will, "An Air Force 'Way of Swarm': Using Wargaming and Artificial Intelligence to Train Drones," War on the Rocks, September 21, 2019.

Stone, Adam, "Army Logistics Integrating New AI, Cloud Capabilities," C4ISRNet, September 7, 2017. As of September 8, 2020:
https://www.c4isrnet.com/home/2017/09/07/
army-logistics-integrating-new-ai-cloud-capabilities/

Sukman, Daniel, "The Institutional Level of War," Strategy Bridge, May 5, 2016. As of July 1, 2021:
https://thestrategybridge.org/the-bridge/2016/5/5/
the-institutional-level-of-war

U.S. Department of Defense, "DOD Announces $250M to Ukraine,"press release, June 11, 2020. As of September 8, 2020:
https://www.defense.gov/Newsroom/Releases/Release/Article/2215888/
dod-announces-250m-to-ukraine/

van den Bosch, Karel, and Adelbert Bronkhorst, "Human-AI Cooperation to Benefit Military Decision Making," paper presented at the Big Data and Artificial Intelligence for Military Decision Making conference, Bordeaux, France, May 30–June 1, 2018. As of September 8, 2020:
https://www.sto.nato.int/publications/STO%20Meeting%20Proceedings/
STO-MP-IST-160/MP-IST-160-S3-1.pdf

Vreeland, Hans, "Targeting the Islamic State, or Why the Military Should Invest in Artificial Intelligence," War on the Rocks, May 16, 2019.

Winn, Zach, "A Human-Machine Collaboration to Defend Against Cyberattacks," MIT News, February 21, 2020. As of September 6, 2020:
https://news.mit.edu/2020/patternex-machine-learning-cybersecurity-0221

Work, Robert, "Remarks by Defense Deputy Secretary Robert Work at the CNAS Inaugural National Security Forum," Center for a New American Security, December 14, 2015.

Wolters, Tod D., "USEUCOM 2020 Posture Statement," testimony before the Senate Armed Services Committee, February 25, 2020. As of September 2014:
https://www.eucom.mil/document/40291/
general-wolters-fy2021-testimony-to-the-senat